LIFE, HEALTH,

AND

DISEASE.

BY EDWARD JOHNSON, M.D.

AUTHOR OF "HYDROPATHY," "NUCES PHILOSOPHICÆ," ETC.

"The pith of nearly all that has been written on the prevention of disease might be included under two heads, almost in two words—TEMPERANCE and EXERCISE."—*Dr. James Johnson.*

NEW YORK:

JOHN WILEY, 161 BROADWAY,

AND 13 PATERNOSTER ROW, LONDON.

1850.

CONTENTS.

LETTER I.

LETTER II.

LETTER III.

LETTER IV.

LETTER V.

LETTER VI.

LETTER VII.

LETTER VIII.

LETTER IX.

LETTER X.

LETTER XI.

PREFACE TO THE NINTH EDITION.

SINCE the publication of the earlier editions of this work, several important discoveries have been made in the sciences of physiology, animal chemistry, and minute anatomy.

The necessary corrections, consequent upon these new discoveries, have been made in this edition. These, however, do not, in the slightest degree, affect any of the great general principles laid down in this work—excepting as so many additional proofs of their accuracy. The doctrines indeed lately promulgated by Liebig are singularly in accordance with, and confirmatory of, all that I have here said on the subject of alcoholic stimulants, as well as on the subject of the causes *of health and strength.* The following passages will put this in a striking light. "The cause of *waste of matter* is the chemical action of oxygen. The act of waste of matter is called the change of matter; it occurs in consequence of the absorption of oxygen into the substance of living parts." "By the absorption of oxygen into the substance of the living tissues, these lose their condition of life, and are separated as lifeless, unorganized (disorganized) compounds." "All experience proves that there is, in the organism, only one source of mechanical power; and this source is the conversion of living parts into lifeless, amorphous compounds"—(in plain language, by the waste of the old body). "Proceeding from this truth, which is independent of all theory, animal life may be viewed as determined by the mutual action of opposed forces; of which one class may be considered as *causes of increase* (of supply of matter) and the other as *causes of diminution* (of waste of matter)." "All vital activity arises from the mutual action of the oxygen of the atmosphere and the elements of the food"—that is, waste and supply.

"That condition of the body which is called health includes the conception of an equilibrium among all the causes of *waste and supply;* and thus animal life is recognized as the mutual action of both; and appears as an alternating destruction and restoration of the state of equilibrium."* Thus Liebig makes *waste*—the disorganization of the body—to be the first of the two causes which constitute the sole source of all vital activity and all animal mechanical power. The first cause is waste,

* Liebig's Animal Chemistry.

the second, supply. Precisely the same doctrines are taught in the following passages from "Life, Health, and Disease." "Health and strength depend upon energetic contractility—energetic contractility depends upon rapid re-organization—rapid re-organization depends upon rapid disorganization—therefore health and strength *depend* upon rapid *disorganization*"—that is, waste of matter.

Again: "The operation by which life is supported may be illustrated by the operation by which motion is supported and communicated by two cog-wheels acting upon each other. Keep your eye steadily fixed upon the point at which the cogs of the two are interlocked. What do you observe? Why, that at every instant, the empty space which is presented by one wheel is instantly filled by a tooth or cog of the other wheel, to be almost immediately emptied again, and again refilled. Thus it is that at every point of the body, and at every instant, little empty spaces are made which are immediately filled by the nutritive arteries, to be again emptied and again filled." This is intended to illustrate the mode in which the all-important process of waste and supply—"sole source of all vital activity, and all mechanical power" in the body—is carried on. And seeing that this process depends upon the presence of oxygen in the ultimate tissues of the organism, and that oxygen can only be carried thither through the medium of an active and vigorous circulation; and seeing further that the circulation can only be impelled to activity and vigor by virtue of bodily exertion, the immense importance of exercise as a hygienic and remedial agent, must be apparent to the most superficial thinker.

With regard to Liebig's opinion of the evil influence of alcohol on the human system, it is extremely gratifying to me to observe that they also are confirmative of my own views on the same subject. After dwelling, in various passages, on the fundamental importance of the "change of matter," as being the sole source of health and strength, he declares that alcohol has the direct effect of putting a stop to the "change of matter." "It is, consequently, obvious," says he, "that by the use of alcohol a limit must rapidly be put to the change of matter in certain parts of the body. The oxygen of the arterial blood which, in the absence of alcohol, would have combined with the matter of the tissues, or with that formed by the metamorphosis of these tissues, now combines with the elements of alcohol. The arterial blood becomes venous, without the substance of the muscles having taken any share in the transformation."

In the new view which I have taken of the source and mode of formation of the alvine evacuations, I am also supported by the doctrines of Liebig. E. J.

Umberslade Hall,
near Birmingham.

PREFACE TO THE FIRST EDITION.

SOMETIME since I received a letter from a very near relative, in which he stated, that he had been, for a considerable time, an invalid, laboring under a combination of the most dissimilar symptoms, all of which, he was assured, are represented by the term "indigestion." He complained bitterly that he could not obtain any satisfactory information as to the real nature of this Protean malady; nor as to the probability of his perfect and permanent recovery. When he questioned his medical attendants on the subject, they evinced every disposition to satisfy him; but they could not avoid making use of phrases which were to him words without meaning. These phrases were, nevertheless, only such as are in common and daily use among all classes—phrases which he himself had frequently used, and supposed he fully understood; but which he now found, when he strictly examined them, really conveyed no definite idea to his mind. He was told that his "digestion was impaired." He asked, what was meant by that; and was told, his "digestive apparatus was deranged in its economy." My brother was still no nearer the mark; and his medical attendant, observing his puzzled look, proceeded to explain and make the matter perfectly clear, by telling him that his "secretions were depraved, his gastric juice deficient, his nutritive functions feebly performed, and that the tone, the energy, the nisus formativus—in fact, the vis vitæ—was fully twenty per cent. below par." The enlightened patient bowed his gratitude for this luminous explanation, and sadly re-seated himself in his chair of sickness—as wise, perhaps, but certainly no wiser, than he was before.

Now, my brother is neither a profound scholar in mathematics, nor proficient in astronomy, nor an adept in chemistry; but he possesses what may be called a gentlemanly acquaintance with all these. That is to say, he understands the great general and fundamental laws which govern them; and therefore, if he were asked a question in any one of these sciences, although he might not be able to answer it, yet he would readily understand its nature and purport; and if the problem involved in that question were explained to him, he would have no difficulty in comprehending it. But not so with the science of health and disease, or rather that which teaches the nature of health and disease:—and the reason clearly is, because he knows nothing whatever of the fundamental laws upon which life depends—nothing whatever of the several actions which constitute life—nothing whatever of the intimate structure of the living organs. He knows neither how he lives, nor how he moves, nor how he breathes. The very language of the science is a dead letter to him; being borrowed, I verily believe, from every tongue that was ever spoken, and, for aught I know, from some that were never spoken at all. It is true he has some notion of the general appearance of a few of the larger organs, because he has seen similar organs in dead animals. For instance, he has seen a great reddish mass of flesh, and has heard it called a liver; and he has been told, that in the liver the bile is made; but beyond this vague and meagre notion he knows literally nothing. As to how it is made—*by what*, and *from what*—or what are the several steps and stages of the operation—as to all this, which constitutes the very kernel of the nut, and without which the shell is nought, he is in worse Cimmerian darkness.

Having read my brother's letter, and digested its contents, I was forcibly struck with the truth of his remarks, and felt that he really had just cause for lamentation. I then conceived that a small concise work, clearly explaining, in common language, the nature of the animal economy—the mechanics of the internal man—the mechanism of life—detailing, step by step, what actually takes place in the performance of each of the functions concerned in the preservation of life and health, and how, and by what causes, life is sustained—it struck me, I say, that such a work would be highly acceptable to the public, and would supply a desideratum in the elementary scientific literature of the country.

There is no mystery into which mankind are more curious to pry than into that of their internal structure;—and certainly there is none on earth which so nearly concerns them. There are many books written with a view to give men a general notion of the laws respecting their property; and it seems to me astonishing that there should not be one calculated to inform them concerning those infinitely more important laws which concern their health. Every gentleman is supposed to know something—the general principles, at least—of all the liberal sciences, excepting that particular one in which alone he has any really great and personal interest. I do think that such a work as I have attempted to describe (providing it were well and plainly written, and all technicalities and unnecessary minutiæ carefully avoided) would be read with great attention and interest, and, I trust, profit by all classes. It would be read, I think, by invalids, in order to acquire some notion as to their own maladies, and so be better qualified to understand and practise the rules of regimen prescribed by their medical advisers: and it would occasionally be consulted, perhaps, by those who were not invalids, in order to acquaint themselves with the best manner of preserving the blessing of which they were then in possession. I believe, also, that such a work would tend more than any other to induce men to practise those rules of conduct which are best calculated to preserve and promote health; because men are ever more ready to do this or that, when they can themselves clearly see and understand its necessity—and the manner how, and reason why, that necessity exists—than when they have no other authority than the dictum of another, however high their respect for his knowledge and judgment may be. Neither, as I think, should medical men take offence at the publication of such a work; as it would have a direct tendency to ennoble their profession—to render it purely scientific—and to divest it of that mystification by which it was formerly so much disgraced, and of which a portion still remains. If patients themselves had a clear general acquaintance with their own internal machinery—with the nature of the several offices intended to be fulfilled by the several parts of that machinery, and with the nature of disease in general; and if, with their mind's eye, they were enabled to look into themselves, and behold the complicated and delicate clock-work—every wheel in motion, every spring in operation—all acting in concert, and all tending to one purpose, yet requiring only the slightest imprudent interference to throw the whole into disorder and irreparable confusion—if, I say, they could see all this, they could not but feel and acknowledge that so beautiful, complicated, and wonderful a machine could only be regulated by the hand of a mechanician intimately acquainted with its minutest structure, and with the particular uses and manner of handling the several instruments necessary to rectify whatever derangement may have accidentally befallen it. It would also materially conduce to destroy the predilection of the public mind for quacks and quackery: for who that knows anything at all of the animal economy, and of the nature of disease, can for one instant be gulled into a belief that any one remedy can be, at all times, good or proper, even for the same disease, and for the same patient? A bumper of brandy will cure the headache, providing that it be caused by a disordered stomach; but a glass of brandy, administered for a headache arising from inflammation of the brain, would, in all human probability, destroy the patient. And how is the patient to know from which of these causes his headache arises?

Such a work I have, to the best of my ability, executed—and I have done so in a series of familiar Letters, because I thought that it would afford me the best opportunity of employing a plain and conversational style, which is the more necessary when writing for readers who have no acquaintance whatever with the subject treated. For the same reason, I have avoided all professional pedantries and learned technicalities, whenever it could be done consistently with perspicuity; and I have described just so much, and no more, of the structure of the body as I thought sufficient to give the general reader a clear idea of those parts only which are concerned in the preservation of health. Thus, in speaking of the heart, I have divided it into two cavities—a right and a left; although, in fact, each of those cavities is again divided into two others. But, as a knowledge of this fact is not at all necessary to the understanding the general functions of the heart; and, as the description of this second division into cavities would necessarily involve another description of mitral, semilunar, and tricuspid valves, fleshy columns, tendinous cords, curtains, &c.—all of which would be " caviare to the general:"—I have thought it best to confine myself to the first grand division—the only one necessary to be known, in order to acquire a lucid notion of the course pursued by the blood. I pretend not herein to teach the anatomical structure of our organs generally; but only to exhibit the several changes necessary to nutrition, which are wrought upon our food within the *ultimate tissue* of those organs—and to show how those changes are effected.

Such are the nature and objects of the work which I now present to the public. Whether I have succeeded in achieving those objects, or not, must be left to the decision of the reader.

LIFE, HEALTH, AND DISEASE.

LETTER I.

Umberslade Hall, near Birmingham.

MY DEAR JOHN,

In reply to your Letter, wherein you complain that you cannot gather any clear notion of the nature of your malady, because you cannot attach, in your own mind, any distinct idea to the terms which your medical attendants seem obliged to use in their endeavors to explain it to you; I am about to give you, in a series of Letters, a plain and familiar description of the mechanism of your internal man; together with a brief history of those internal motions and actions which constitute animal life, and any disturbance in the harmony of which constitutes disease. Thus, I think, I shall enable you easily to surmount the difficulty of which you complain. There is another benefit which I intend should result from these Letters. They will, I hope, enable you to understand what are the habits of life which are most likely to conduce to a sound mind and a sound body. For if I were requested to teach a man how to regulate and repair his watch so as to make it keep true time, I should think the best way to enable him to do this, would be to make him acquainted with the internal mechanism of a watch —showing him the uses of the several wheels and springs which keep his watch going. So, I believe, the best plan to teach men how to regulate their diet and habits of life, so as to make their health keep true time, is to make them acquainted with the mechanism of their internal selves—showing them the uses of the several organs and fluids which keep *life* going.

But, before we descend to particulars, it will be as well to take a rapid and brief, but general survey of the several parts which go to the composition of the animal, man. I say, the animal—because here we have nothing whatever to do with the higher attributes of his nature*—attributes which have no connection with physical structure, and the phenomena of which are wholly independent of all physical laws. We are here wholly and solely concerned simply with the physical animal.

The method I shall adopt, in order to exhibit the principal systems of which the whole scheme of man is made up, and to show the relation which exists between them, and the dependence of one upon another, may be considered as fanciful. Perhaps it is so. But it struck me as one well calculated to render what I wish to say, easily comprehensible; and that circumstance alone is a good recommendation: for I am not ambitious of fine writing, either as it regards accurate arrangement, philosophical speculation, or learned and elegant diction. I am only anxious to be understood.

If man had been the work of any being less than Omniscient, the several single ideas composing the one complex idea of man must have occurred in succession; and the first must have been the idea of his figure. The first idea could only have been, as I shall presently prove, merely that of an image or statue of the particular form and appearance which man presents. I am, of course, for the present, supposing man to have been the first animal produced, and that his artificer was some being of inferior wisdom to that of Him who is, in truth, his real Author.

Having conceived the idea of a particular figure, and determined to realize it, the next point to be settled would be the kind of materials of which to fashion it. Having chosen bone, and shaped his image according to his preconceived idea, the first of a series of single ideas forming the one complex idea would be realized; and a solid statue of bone would have been the result—a mere image of the human form.

Contemplating the work of his hands, the desire of endowing it with powers of locomotion might then occur to him. In order to accomplish this, the artificer would find it necessary, first, to divide the statue into parts (re-uniting these parts by means of joints), and then to contrive a number of motive instruments, which, being attached to the jointed statue, would enable it to move—as the mechanic who wishes to move a heavy weight must first construct his instruments of motion, such as wheels, pulleys, levers, &c. Having effected this contrivance, the second idea of the series would be realized—the idea of the muscular system.

* By which I mean—not the mind—not the reasoning faculties; but the soul, its immortality, &c.

But when he had contrived and attached his muscles, he would find that the particular shape and general appearance, which he had predetermined his work should bear, had been quite destroyed, and that these same muscles attached to the outside of the statue were a terrible disfigurement of its external beauty and symmetry. To remedy this evil, it would be necessary to scoop and pare down, and hollow out, different parts of the image, and then to fill up these hollows with his muscles; and thus restore those parts, which had been so cut down, to their original size, and again bring his image to its former shape and dimensions, by taking away a bulk of bone equal to the bulk of muscle which he wanted to add. But still he would find, notwithstanding his muscles, that his statue could not yet move, any more than a steam-engine can move merely because it has wheels, unless there be some power to set those wheels in motion. Hence would arise the third idea of the series—that of a nervous system, whose office it is to afford motive power to the muscles, which are of themselves only motive instruments. This motive power is to the muscles—which are, in fact, only so many pulleys, ropes, &c.—what the mechanic's hand is to the pulleys, wheels, &c.: it sets them in motion, and keeps them moving. And here, again, he would be obliged to hollow out another portion of the bone, in order to make room for the brain and spinal marrow (from which nearly all the nerves arise), so that their attachment might not destroy the symmetry of his image. The nerves which arise from the brain and spinal marrow, and whose office it is to carry the motive power to the muscles, he would of course distribute and conceal among the numberless little bundles of fibres of which the muscles are composed. And this he might easily do, seeing that the nerves are merely small threads, and therefore easily concealed and embedded in the soft parts, without producing any disfigurement or much apparent increase of bulk.

Again contemplating his production, it would occur to him, that the materials of which he had found it necessary to construct it, were liable to decomposition and decay—putrefaction. To surmount this new difficulty, it would be incumbent upon him to contrive a conservative system: and hence he would arrive at the fourth idea—that of a system of nutrition. As the organs of this system are large and numerous, he would be compelled to hollow out the whole body of the statue, in order to make room for them, and put them out of sight; leaving no more of solid image than just sufficient to support and give attachment to the several new contrivances which, in improving upon his original idea, he had been obliged to add.

Once more contemplating his work, he would now be delighted to see his new, animated, and improved statue moving from place to place, without assistance. His satisfaction, however, would be somewhat disturbed, by observing the grotesque, awkward, and uncertain manner in which it proceeded or rather zigzagged; and very soon all his joy would be suddenly turned into consternation, by beholding his unhappy automaton all at once break its head against a post, or hop into a river, and vanish beneath the waters.

Having fished it up from the stream, or mended its broken head, it would now be tolerably clear to him that his new creation was not yet perfect. He would see that it was absolutely necessary to its safety that it should know when its path was obstructed by a post or a pond. This would suggest the idea of the organs of the senses; being the fifth idea, and completing that series, of which the complex one, represented by the words "animal man," is composed. By the organs of the senses, his object would be, to establish a certain relation between it and the rest of the world—to enable it to acquire ideas (by means of the experiences of these senses) of whatever was likely to inflict injury or afford pleasure, that it might seek the one and avoid the other.

In considering what senses were necessary, he would find that five were required. Having scooped away another portion of what little of the bony statue yet remained, and so introduced the eye and ear; and having found proper places for the addition of the organs of taste and smell; and thus, having disposed of four out of the five senses required, he might be supposed to pause, from suddenly observing that there was yet an imperfection which had escaped his notice: for he would see that the external surface of his image was very unequal, from the many scoopings and hollowings which it had undergone—that, though these had been filled up by muscles, &c., they did not fit with sufficient accuracy to make all smooth—that some parts were soft, and others hard—that some were of one color, and some of another;—whereas the image, according to his preconceived idea, should have been, as to its external appearance, everywhere even, homogeneous, and soft to the touch. How was this imperfection to be rectified? Having still the fifth sense to add, he would resolve, it may be supposed, to make use of this sense to restore the image to its originally predetermined external appearance of homogeneous beauty. Instead, therefore, of making the sense of touch reside in a single organ, he spread it over the entire surface by means of a most delicate and beautiful membrane (the skin); and wrapping the image in it, and seeing that it fitted with perfect accuracy, he would be gratified to find that he had killed two birds with one stone—that he had given to his image the desired appearance—and, at the same time, found the means of attaching the fifth sense without any more hollowing and scooping.

But these organs of the senses being of themselves only instruments (as the muscles of themselves are only instruments of motion), could be of no use, unless they were supplied by some other organ with the power of sensation. But he could find no room for any more organs, nor were more necessary; for he had only to take care that certain of the nerves should be distributed to them, and his organs of sense would be thus at once supplied with the power of perfectly fulfilling their respective offices, by means of a sensibility supplied to them by these nerves. So, again, some of the organs of nutrition, of which the heart is one, would have been useless, unless he had taken care that they also were properly supplied, by means of the nerves, with motive power.

Having contrived all this to his satisfaction, he once again, it may be supposed, carefully reviewed his work;—and having tipped the fingers and toes with nails, ornamented and protected the head and chin with hair, pencilled the eyebrows, and fringed the eyelids; and having furnished his jaws with teeth, and taught him how to use them; he would be glad that he had finished his work; for, I think, he must have found it a very troublesome affair.

I do not pretend to say that the several ideas composing the series would have occurred in the particular order in which I have arranged them. I only say, they must have arisen one after another; and that of the mere image must have been the first; for, surely, it seems absurd to suppose that the idea of a peculiar set of instruments of motion could occur before the idea of something peculiar to be moved—that the idea of a peculiar system of nutrition should present itself before the idea of a peculiar something to be nourished—the idea of a peculiar source of motive power before the idea of a something to be put in motion;—or that the idea of a peculiar system of organs of sense, which may justly be termed organs for the admission of pleasure and avoidance of pain, should precede the idea of something to be pained or pleased.

Besides, all the other parts of the body are subservient to the mere image or skeleton; and we can assign a distinct purpose to every part of the body except this same skeleton. Thus, if one be asked, Why the muscular system exists? the answer is ready—to move the bones. Why the nutritive system?—to enable the bony edifice, together with its motive appendages, to resist decay:—and so of all the other systems. But if it be asked, Why was the skeleton created?—who shall answer the question? Its purpose cannot be simply to support the soft parts—this is only its secondary and adventitious use: for I have just shown that the soft parts are subservient to it, and not it to the soft parts. And, besides, I have proved that the idea of the bony image must of necessity have been the first to present itself, since all the other parts of the body answer no other purpose than that of conferring some service upon the skeleton or bony image.

To the Divine Artificer, however, the source of all wisdom, and fountain of all that is good and all that is beautiful, and to whom nothing is impossible, nothing difficult—to Him the whole complex idea was present at the same instant of time: and therefore foreseeing, in his omniscience, all that which could only occur to a finite mind step by step, and little by little, He from the first provided against all exigencies, and fashioned man at once as he is.

Thus, then, we see that the body may be conveniently (at least for our present purpose) divided into five principal systems. First come the bones: secondly, the muscles moving these: thirdly, the nervous system, from which the motive power of the muscles is derived: fourthly, the organs of the senses, by which the being, thus endowed, discriminates between that which is conducive, and that which is detrimental, to his welfare; enabling him to seek the former and shun the latter; and establishing a due connection between him and the objects which surround him: and, fifthly, the nutritive organs, nourishing and sustaining life in the whole.

Now what is my motive for dividing the body into its several systems, and for thus carefully showing you that all the other systems composing the body are only so many servants in waiting upon the bony fabric or skeleton? It is a very important one. It is this. It is in order to show you that the only difference between the animal man and other animals of the higher order, is to be found in the shape of his skeleton. It is the skeleton which constitutes and individualizes the animal. For all animals (at least of the highest order, and living in the same element) have, equally with men, the same kind of muscular system, the same kind of nervous system, the same kind of nutritive system, and similar organs of the senses—and all these are, equally with those of man, subservient to the well-being and duration of the skeleton. They are different, it is true, in size and figure, but this is only in consequence of the difference in size and figure of the skeleton to which they are attached. I am particularly anxious that you should see and understand all this: because you will then be enabled to appreciate the force of those arguments, relative to the best manner of promoting the health and strength of man, which are drawn from the observation of those means which are acknowledged to be the best for promoting the health and strength of certain animals, such as the horse, the dog, &c., &c. You will then be able plainly to perceive, that if water be the best drink for a horse, so also it must necessarily—I say *necessarily*—be for man. Why? I have just told you. Because both animals are the same in physical constitution. The only reason which can be given why water and a vegetable diet are most conducive to the health of the horse, is because these are his *natural* drink and his *natural* diet. And seeing that there is no essential difference between the living economy of the animal horse and the animal man—seeing that, in this respect, they both belong to one class—it follows that if the natural drink of the one be the best for that one, so the natural drink of the other must be best for that other. For they are but two individuals of one class. Whatever is true of a class is true of each individual of that class. If, therefore, it be true that the drink which nature has provided for animals is the best they can have, it is equally true that the drink which she has provided for man (water) is also the best which *he* can have—he being (in all that concerns his physical constitution) but a different individual of the same class. If nature had provided wine for man, then wine would have been (necessarily) the best possible drink for man, for the very same reason that water is the best possible drink for horses. If there had been any intrinsic difference between them—if the living actions of the one had been different from those of the other—it might then have been said that nature, in pity to the very small degree of art which the brute is capable of acquiring, took care to leave nothing to be accomplished by art. But that, foreseeing the very high degree of art which man, through his faculty of speech, would arrive at, she did, in his case, leave something to be artificially effected. It might have been said: "It is true that for those animals whose life consists simply in the due performance of the functions of absorption, secretion, circulation, and respiration, the drink furnished by nature is all-sufficient. But here—in man—there is another function of life, which the brute has not, and in order that this other function may be duly performed, it is necessary to have recourse to art. Thus much—the care of this particular function, distinctive and characteristic of human life only—nature has left to be provided for by the wisdom and ingenuity of men." But I say there is no such distinguishing and characteristic function in human life.

But you will say—"Does not man, in his present artificial state, require a different drink—a stimulating one?" Most manifestly not—seeing that the very evil entailed upon us by our artificial state, is one of excessive stimulation—excessive cerebral excitement; of which that host of diseases, consisting of vascular congestion, is the direct result. We are constantly blowing the fire all day long, and when the bellows are laid aside, during the night, or any other interval, the fire burns so dully that it is incapable of affording heat enough to perform the proper functions of a fire. Then we seize the bellows and begin to blow again, and thus we keep up a flickering and uncertain flame. But to insure a steady and efficient fire, the right way is, to rake out the dead ashes and cinders every now and then, to supply it lightly at proper intervals with good and proper fuel, and leave the rest to the natural draught of the chimney. The cook who is constantly blowing the fire will only waste coals, and will never have a good one.

The life of cultivated society is one of hourly excitement—the excitements of business—of emulation—of ambition—of reading—of writing—of quarrelling, &c., &c. The brain is constantly erect—throbbing with thought. What it wants is, quiet—time to recover its exhausted energies. If a man feel languid—dull spirited—enfeebled—it is because his energies have been overtasked. In such cases the remedy is rest—not still further stimulation.

And you will further see that if exercise in the open air be necessary to the health of a horse, so also it must be for that of man. And that if hot stables and warm clothing be injurious to the health and strength of the horse, so also must warm parlors, closed curtains, soft beds, and such other similar luxuries, be equally detrimental to the health and strength of man. It must of *necessity* be so, since the structural nature of the two animals is one and the same, saving only in figure and bulk; and since there is no difference—no, not so much as the estimation of the millionth part of a hair—between the actions which constitute the life of a horse and those which constitute the life of man.

We have now taken a view, in the gross, of the several parts of which man is made up. But, in contemplating the new being, as we have seen him turned out of the hands of our imaginary Prometheus, has it not struck you, that something extremely necessary to his safety has been forgotten? Let us suppose that the first living thing he meets, after his creation, is a wolf, gaunt with hunger. He cannot flee from him; for the wolf is swifter than he: he cannot resist him; for the wolf is stronger than he: he must perish, for want of weapons of defence. In vain do his organs of the senses warn him of the approach of danger: he can neither shun it, nor resist it. What he still wants, then, to secure him in his existence, are weapons of offence and defence.

Here, then, is a new difficulty. How is it to be overcome? Man was originally designed, not only not to become the prey of the rest of the animal creation, but to hold every other animal under his own dominion—within his own power—under his own control, and at his own service. Nothing short of Omnipotence could effect it; and I know of no better proof of the divine origin of man, than the solution of the following problem:—man and the beast of prey being given, to give to the weaker dominion over the stronger. How beautifully, perfectly, yet simply, has the Almighty Ruler surmounted this difficulty. Man *speaks!*—and the problem is solved. By virtue of that miraculous little system of organs—the organs of speech—what an immeasurable distance is, at one instant, interposed between the reasoning powers of the brute and his own! These enable him to add to the stock of knowledge resulting from his own experience, the whole stock acquired by the experience of his fellow-man. Thus he obtains innumerable ideas, which could never have been collected by the senses of one individual. These he combines, analyzes, recombines, compares, arranges. New ideas give rise to new pursuits, and new pursuits to new ideas. Thus his stock of knowledge is continually augmented, as his sources of ideas are multiplied; till his power, resulting from his knowledge, is only inferior to that of his Creator. The lion is lord of the forest; but man is lord of the lion. The stag and the antelope outstrip the wind; but man outstrips the antelope and stag. The most powerful of the brute creation become his obedient slaves. The tiger is hunted and slain, or entrapped and imprisoned, and his savage ferocity made subservient to his master's amusement and profit. Man's superior reason, therefore—which he owes chiefly to his faculty of speech—constitutes his weapon of offence and defence.

Φυσις κερατα ταυροις,
Ὁπλας δ' εδωκεν ἱπποις,
Ποδωκιην λαγωοις,
Λεουσι χασμ' οδοντων,
Τοις ιχθυσιν το νηκτον,
Τοις ορνεοις πετασθαι,
Τοις ανδρασι ΦΡΟΝΗΜΑ.

"Nature hath given (for their weapons of offence and defence) horns to bulls, hoofs to horses, swiftness to hares, a cavern of teeth to lions; to fishes, the power of swimming; to birds, the power of flying;—to man she has given *wisdom.*"

"Man," says Geoffrey Crayon, "is naturally more prone to subtlety than open valor, owing to his physical weakness, in comparison with other animals. They are endowed with natural weapons of defence—with horns, with tusks, with hoofs, and talons; but man has to depend on his superior sagacity."

I hope it is not necessary to tell you, who know me so well, that I consider the reasoning faculty as quite distinct from the soul, which I believe to be a portion of the divine essence, "divinam particulam auræ," inhabiting the body, but not subservient to any of its functions—Ου λογος, αλλα τι κρειττον. "Not reason, but something better."

I have mentioned the reasoning powers of brutes. No one, I think, of the present day, who is accustomed to read, and think, and take note of the habits of animals, will deny their possession of this faculty. Everything which remembers, and regulates its conduct by this remembrance, performs an act of reason. Why should they not reason? And that man owes his superiority of reasoning power to his faculty of speech, is most strikingly and irresistibly proved by the effect of the press. What is printing, but an extension of the powers of speaking? Enabling a man, without moving from his native soil, to put his antipodes in possession of every new idea he acquires; so that what one acquires is acquired by all; thus multiplying the still newer ideas to which this newly-acquired one may give rise, by nearly the whole number of the reading inhabitants of the world: for almost every man will probably derive, from the combination of this new idea with one which he already possesses, another new idea;—and this other new idea is again told to the world through the press, and its results again multiplied as before. The first possession of the faculty of speech did not elevate man nearly so much above the brute, as this extension of it has lifted him above his former self. It must be remembered, also, that it is solely owing to this faculty of speech that men can *consult together*—an important advantage which is denied to brutes.

It may be objected that man has a larger brain than other animals, and that his superior ratiocinative powers may be owing to this circumstance. The objection may be answered in two ways. But I shall only answer it as though it were true; for if it be true, it does not invalidate my argument: for if man possess a larger brain, it is only in consequence of his possessing the organs of speech. Because, that man should speak was a part of his original design; and the Creator, foreknowing (as he foreknew and provided for every other exigency) that the faculty of speech would render a larger brain necessary for the reception of that multitudinous host of ideas which his vocal organs would enable him to muster—and in order that he might reap the full advantage which His gift of speech was calculated to confer upon him—has given him a magnitude of brain corresponding to his necessity. If he had not done so, he would have defeated his own purpose: he would have given him the means of acquiring ideas, without the means of turning them to account; and man, as it regards his reason, would still have been but one remove above his neighbor, the brute. His superior magnitude of brain, therefore (if he possess it), and his superior ratiocinative faculties, are both alike the consequence of his vocal organs. If it be true that he possess a larger brain than other animals, this may be a *second* cause of his superior reason; but his possession of speech still remains the *first* cause.

But it is manifested that the power of acquiring knowledge is not in proportion to mere magnitude of brain, for my lady's poodle is just as sagacious (if not more so) as my lord's coach-horse. And sagacity is but another name for knowledge.

There is a difference between the human brain and that of the brute. If an animal, having a larger brain than man, were endowed with the faculty of speech, although it would lift his ratiocinative powers to an elevation nearly equal to the grandeur of man's, it would not quite equal it. But this difference is to be sought for, not so much in superior magnitude, as in superior delicacy, elaborateness, and intricacy of structure—in the greater quantity of cineritious matter, and in the greater size and number of its convolutions. As this superior quality, like the supposed superior magnitude of man's brain, is only the consequence of his possessing articulating organs, my assertion still holds, that his enunciatory apparatus is the sole cause of his superior ratiocinative capabilities.

Now, my dear John, begging pardon for this long digression about the talking organs, and with a devout hope that your amiable wife, when she learns how much she owes to these little instruments, will be particularly careful never, by overtasking, to put them out of tune, and you out of temper, I will descend from generals to particulars.

Many years ago, it was believed, by physicians, that our food was operated on by the stomach, pretty much in the same way as shins of beef and ox-cheeks are dealt with by Papin's digester. It was supposed to be digested: that is, simmered, concocted, or stewed. When a man, therefore, felt himself strong and active, not oppressed after meals, and altogether in excellent health and spirits, this fine state of things was thought to be all owing to the circumstance of the stewing and simmering of the stomach having been carried on merrily and well, till it was done enough; and then, it was thought, the stomach handed over the stew to the bowels, thoroughly and properly cooked. But when a man without any one very painful symptom in particular, felt himself generally indisposed, weakly, disinclined to action, low-spirited, and oppressed after eating; it was then said that his food had not been properly digested—in plain English, not properly stewed by the stomach; but that it had been left by that organ very much in the same state in which the shins of beef would be found after having been stewed over a bad fire, and in a cracked digester which let out the steam. He was said to be afflicted with indigestion! which signifies the unequal distribution of particles by stewing, or simply, imperfect stewing. Or, if his physician chanced to be somewhat of a pedant, the more learned word dyspepsia was used, which signifies difficult boiling.

You see, therefore, that when these queer words, digestion, indigestion, dyspepsia, digestive, &c., &c., were first introduced, viz., when physicians looked upon the stomach as little more than a living stew-pan, they had each a very distinct and definite meaning, and were used with perfect propriety. I mention this merely to account to you for the introduction of these strange words into medical language. That these words are still used by medical men is of little consequence; because, although they retain the old words, they attach to them new meanings—meanings which by no means belong to the words, but which are perfectly understood among themselves. But with the rest of mankind the case is very different: for, as they retain the old words, they must also retain the old meanings, or else no meaning at all, which is by far most frequently the case; because they cannot be aware of the several great changes and improvements which medical philosophy has undergone. When these words, therefore, were first introduced, they were proper enough; but now that physicians have discarded "Papin's digester," and refuse to recognize any similarity between the uses of the stomach and those of the stew-pan—now that we know that the "stomach is neither a mill, nor a stew-pan, nor a fermenting vat, but a stomach, gentlemen, a stomach"—now, I say, all these words, as applied to any condition or action of any part of the body, are perfectly senseless and worse than absurd, because they are only productive of confusion—they ought, therefore, to be expelled from medical language; or if retained, they must, and indeed can only be understood in senses which do not properly belong to them. But we had better get rid of them altogether; for it is impossible to use them with the slightest shadow of propriety; and we shall have no difficulty in finding substitutes, each of which will carry its own definite and obvious signification.

From what I have said of the manner and reason of the introduction of these words, which we have just ejected from our vocabulary, you will easily understand how it came to pass that all those disorders to which the term indigestion is applied, were supposed to exist in the stomach only; because you will have observed that it was in the stomach, according to the creed of our good forefathers, that all the stewing and digesting were carried on; and when the stew was not properly stewed, they never thought of looking to the fire or to the cook-maid for the cause; they only looked to the stew-pan, which they learnedly denominated "the digestive or stewing organ."

Now for our substitutes.—For the phrase "sound digestion" substitute perfect assimilation; for "indigestion" substitute imperfect assimilation; assimilating organs for "digestive organs:"

assimilation of food instead of "digestion of food," &c., &c. The word assimilation is generally used by authors to designate that process by which the food, after having undergone all the necessary previous changes, ceases to be food, and becomes part and parcel of the living body; when that which was flesh of the dead ox becomes flesh of the living man, or bone, or hair, or skin, &c., &c., according to the nature of the different parts of the body to which it is distributed for the purpose of being assimilated. But, in fact, all the changes which the food undergoes are assimilating changes, all tending to that ultimate assimilation which converts the fluid blood into the solid body—in one word, its solidification. Thus the conversion of food in the stomach into chyme (learnedly called chymification) is its assimilation to the nature of chyme: its conversion into chyle (chylification) is its assimilation to the nature of chyle: its conversion into blood (sanguinification) is its assimilation to the nature of blood: and if we wish to particularize any one of these changes, we have only to name the organ in which it takes place. Thus, if we wish to designate those particular changes which take place in the stomach, commonly called digestion, we have only to speak of them as "assimilation in the stomach."

Now that we have got rid of the word digestion, with all its stewing derivatives, and banished it from our own to the cook's vocabulary, to which alone it properly belongs, you will not be surprised when I tell you, that the stomach and bowels are by no means the only assimilating organs we possess. Every organ which is concerned in the nutrition of the body—and without a healthy state of which organ, nutrition cannot be properly performed—has a right to be called an assimilating or nutritive organ. I need not tell you that nutrition and assimilation are the same thing. Assimilation completed is nutrition completed; and the several assimilating changes in the food are only so many nutritive steps towards the completion of the process of nutrition.

I shall presently take a meal of food, and trace it, or rather follow it, through all its changes, until it has become assimilated, that is, until it has become part of the living body. In doing this, you will learn what are the assimilating or nutritive organs, what is the office of each, what the changes which these organs severally effect in the food, and in what manner they accomplish these changes.

It will be convenient to state here that two kinds of blood are contained in the body, differing from each other as much as any two things can well differ. The one is of a beautiful, bright, vermilion color, teeming with the living principle, pregnant with all those elements from which all the organs of the body are nourished, and in a condition readily and instantly to part with those elements, each at the proper moment and in the proper place, accordingly as the nutrition of the several parts of the body requires them. This vermilion blood is, as it were, in a state of excitement, being surcharged, not with the principles of electricity, but with the principles of living matter; and, as it circulates through the minute vessels, parts with those living elements with the readiness and freedom with which a highly-excited body parts with its electricity. This blood is conveyed in vessels called arteries. The other kind of blood is a filthy, thick, purplish, blackish, inky puddle, unendowed with any good quality, endowed with many pernicious ones, productive of much mischief, but incapable of any one good with which I am acquainted; save only that from it the bile and the urine are formed. This blood is contained in vessels called veins. Some of the principal differences between arteries and veins are the following:—The arteries carry the living blood from the heart to every point of the body. The veins, like so many waste pipes, carry the deteriorated, dirty, and, if I may so speak, dead and useless blood, from every point of the body, back to the heart. The arteries, arising chiefly by one large trunk from the heart, become smaller and smaller as they pursue their course towards their termination in the veins. The veins, arising from the innumerable terminations of arteries, become larger and larger as they proceed towards the heart. The arteries, therefore, in the neighborhood of the heart—from which, as I have just said, they almost all arise by one common root (the aorta)—are large and few; but from the sides of these there are perpetually given off smaller and smaller branches; while from these smaller, others still smaller than they are continually separating; and so on, until the whole are finally lost in indistinguishable minuteness.

While the arteries are in this state of wonderful attenuation, their course is exceedingly tortuous; they recoil upon themselves; and are circumflexed hither and thither, until there is scarcely a point in the body which is not occupied by one of these little vessels. After having thus permeated the universal body, they lose the characteristics of arteries, and assume the structure of veins.

The terminations of arteries, therefore, are the beginnings of veins. This termination of arteries in veins can be seen, by the aid of the microscope, in the frog and salamander. In some fishes it can be seen with the naked eye. The arteries near their termination, and the veins near their beginning, are many times smaller than the finest hairs; and, in this state of hairlike minuteness, they constitute that which is called the ultimate tissue of the arteries and veins;—and so, also, the tissue formed by the nerves and absorbent vessels, while in their last state of minuteness, is called the ultimate tissue of the nerves and absorbents; and that beautiful network formed by the interlacing of all these delicate and hairlike threads, viz. arteries, veins, nerves, and absorbents, in their minutest condition, is denominated the ultimate tissue of the body; and this ultimate tissue constitutes, in fact, nearly the whole of the body: for all that which appears to our eyes so firm and solid (not even excepting the bones) mainly consists of this astonishing network of minute vessels and nerves. This network, or ultimate tissue of the body, owes its compactness to its being firmly compressed and interwoven; to its being well and accurately filled with fluid (principally blood); and to the circumstances of its being everywhere supported, held together, contained, and, as it were, closely stowed away in the cells of the cellular substance—now more properly called the areolar web, or membrane.

In order to obtain a clear notion of the cellular substance, its universality and appearance, just fancy it possible for an anatomist, with a finely pointed instrument, to pick away every part of your body, which is *not* cellular substance; what remained would be, of course, cellular substance only, and you would present exactly the appearance of a man made of honeycomb or sponge. But if this spongy relique of you were perfectly dried, it would be so light, that the sigh of a butterfly in love would be sufficient to scatter it to the four winds of heaven. Notwithstanding it pervades, therefore, the whole body, its actual quantity or weight is exceedingly small; and notwithstanding its resemblance to sponge when seen in the mass, when spread out and viewed under the microscope, it is observed to present the appearance of a spider's web.

I trust you have now a tolerably accurate idea of the ultimate tissue. If you have not, I pray you to refer back, and read again: and every now and then shut your eyes, and so endeavor to ascertain whether you clearly understand what you have just read or not, and by no means proceed to a second sentence before you have fully understood the first. Pardon this earnestness; and if I have been somewhat tedious or tautological, you must pardon that too, for I am extremely anxious that you should obtain a distinct conception of the nature of this amazing part of our structure; otherwise I shall have lost both my time and labor, and it will be impossible for you to understand me when I come to speak of diet, the conduct of life as it relates to the preservation of health, the origin of disease, &c., all of which have a direct reference to this same ultimate tissue. It is, besides, the most beautiful, the most wonderful, the most important structure in the human fabric, magnificent in its very simplicity, stupendous in its very minuteness; and it is the secret chamber in which Nature conducts all her hidden operations. Hither are brought, and dealt with, by that subtle and mysterious Operator, all the elements necessary to the production of a Newton, or a Montaigne; a Howard, or a Robespierre; a Richard the First of England, or a Lewis the Eleventh of France; a genius, or a dunce; a martyr for religion, or a murderer for pelf.

The physical is the father of the moral man; and it is quite true, "Quod animi mores temperamenta sequantur:"—that morals depend upon temperaments. And no less true is it, that "Philosophy has been in the wrong, not to descend more deeply into the physical man:—there it is that the moral man lies concealed:—the outward man is only the shell of the man within." To alter a man's moral character, you need only alter his physical condition. Take the brave and hardy mountaineer from his hills—lap him in luxury—let him be fed on dainties and couched on down—let his lullaby be sounded by the "soft breathing of the lascivious lute," instead of the wild music of the whistling wind—you will soon reduce him, first physically, and then morally, to the rotund but helpless condition of the turtle-fed, yet imbecile alderman. In a few years replace him on his mountain-top—set him beside his former companions—show him the aggressor against his rights, the oppressor of his race—bid him meet and repel the common enemy. Behold! his courage has fled; the love of liberty and independence is dead within him; the spirit of freedom sleeps; he trembles, and yields, preferring the indolence of slavery to the toil necessary to preserve him free. It may be said, that courage is but one of the moral qualities: true, but it is one on which many others depend. Courage results from a consciousness of physical strength; and cowardice, from a consciousness of physical weakness. The strong will not shun danger, because he feels himself competent to resist and repel it. The weak man, knowing himself unable to surmount danger by an exercise of strength which he does not possess, will resort to other means of preservation—to petty cunning, wily stratagem, mean subterfuge, lying and circumvention. Thus the virtues which are directly opposed to these vices all depend upon courage, at least to a considerable extent; and courage depends on physical strength, the size of the heart and lungs, the firmness of the heart's structural fibre, and the liveliness and energy with which circulation and animalization are performed. The fortitude with which the Indian savage endures torture at the stake, I shall endeavor to show, by and by, is clearly the result of his physical condition. It may be objected that we have numerous instances of undoubted courage in men possessing but little physical strength; but this objection will not hold. When the noble scion of a noble house, the nursling of luxurious ease from his cradle, goes out to fight a duel, is it because he loves danger for the sake of the pleasurable excitement it affords? No.—Is it because he is indifferent to danger? No.—What is it, then, which urges him on? It is the fear of disgrace; it is the dread of being hooted from that sphere of society in which he moves; it is his fear of the finger of scorn which impels him: this, therefore, is not courage—this is fear. If he refuse to fight, he knows that he will be degraded from his caste—his views, whether of love or ambition, will be destroyed. If he fight, he has a chance of escape, and if he escape, his character, as a man of courage, is established. His, therefore, is a choice of two evils; and he chooses to fight as being the less evil of the two. If he could avoid both evils, most assuredly he would do so. But this is not courage. The mere act of fighting does not constitute bravery. It is the feeling, the inward feeling which he carries with him to the field—it is this which constitutes true valor. The rankest coward that ever lived will fight, when he knows that instant death attends his refusal, or that there is more danger in running away than in going forward. True courage loves danger for the sake of the excitement it affords—loves it for the same reason that men love wine—loves it, too, for the glory consequent on overcoming it. Had Richard the First not been the giant he was, would he have been the hero he was? would he have courted danger as he did, alone, and single-handed?

I have said, that many virtues depend on this single quality of courage. Richard possessed the ne plus ultra of courage, and he was high-minded and generous to a fault. He sought to accomplish all his ends openly, avowedly, and honorably, because he felt himself able to do so. His brother John was a coward: and how did he seek to accomplish his objects? Why, by every species of low and cunning villany, not stopping even at murder. Had John been physically constituted as Richard, and Richard as John, John had been called "the lion-hearted," and Richard "the craven coward."

Again, it may be urged, that on the field of battle men not physically strong have frequently performed feats of gratuitous and uncalled-for daring. But neither will this objection serve; for at the time of performing these deeds of valor, their physical constitution is actually altered. The nervous system, powerfully excited by the senses, the trumpet's clang, the panoply of war, the martial music, the stir, the life, the uproar all around, pours into the heart a resistless tide, as it were, of nervous energy; and the heart, obedient to the impulse, propels the blood in a stream of triple force along the arteries, until every organ of the body is in a state of the highest excitement, swollen and distended with the living current. Thus, for a time, the weak become actually strong; and hence these instances of courage in the weak. The same thing occurs in anger. A man under the influence of rage not only appears to possess, but really does possess, triple the physical power which he can command when calm.

It must be remembered that a man may be physically strong, notwithstanding that he may have but a small development of muscle and bone. His brain and spinal cord, his heart, his lungs, may all be largely developed; and this will give him a consciousness of strength—of internal strength—although the *instruments* of strength (muscle and bone)—the instruments to be set in motion by this internal strength, are deficient in size and substance. It is this internal consciousness of living strength which constitutes the difference between the high and low bred horse.

Not only, therefore, is the body constructed in the ultimate tissue, but the character is constructed there also. And as the health and strength of the body depend upon the healthy performance of the processes of assimilation in the ultimate tissue of the body, so also do the health and strength of the character and mind.

> "Who 'd pique himself on intellect, whose use
> Depends so much upon the gastric juice?"

says Byron.

The moral qualities are therefore, at least to a great extent, the offspring of physical structure. I know that moral causes may, and often do, produce physical disease: but this does not weaken the argument; for a child may destroy its parent; and so the moral qualities, though they result from physical structure, may nevertheless re-act upon that structure to its detriment. The qualities of the mind, also, may be modified, improved, trained, and properly directed, by religion and education. So, also, may the child of one parent be nurtured and educated by another.

One of the most familiar instances of the influence of physical conformation on moral character is to be found in the fact, that all the most courageous and ferocious animals have a heart remarkably large and strong in proportion to their size, while the weak and timid have hearts proportionally small. It is as impossible for an animal with a small, flabby heart to be bold and strong, as for two and two to equal five: and equally impossible is it for a man who is physically constructed to be a coward, by any act of his own will, or of abstract courage, to make himself a hero.

I am glad to perceive, by some late publications, that the truth of this doctrine is beginning to be admitted; and I trust it will not be long before parents can be made to understand, that the only certain method of assuring to children a vigorous and healthy mind is, first of all, to allow them the opportunity of acquiring a vigorous and healthy body. Let them be assured, too, that those who begin by cramming a child's memory (for judgment is out of the question) with a quantity of bad French and worse Latin, together with the terms and problems of the abstruse sciences—which, after all, they can only learn to repeat as the parrot does, by rote, without understanding;—let them, I say, be assured, that those who thus begin, by seeking to make a child so very, very learned, will end, in all human probability, by making him sickly.

I have been seduced by this important subject into a long digression:—but let us return to the arteries and veins.

The arteries, ramifying in every direction, like the branches of a tree, from their common root in the heart, and having shot their minute and hairlike terminations into every part of the body, so that you cannot insert the point of the finest needle without wounding one or more of them, cease to be arteries and take the structure of veins. These hairlike veins (which are merely a continuation of hairlike arteries with an alteration in the structure of their coats) soon begin to unite two into one, to form larger veins. These larger veins again presently unite two or more into one, to form larger still, until all the veins of the body have united together, and so formed two very large ones, which empty themselves into the heart. One of the grand distinctions, then, between veins and arteries is, that while the arteries arising from the heart are multiplied in number and diminished in size, until they have reached and distributed their blood to the ultimate tissue, the veins arising from the ultimate tissue are constantly becoming diminished in number and increased in size, until they have reached and carried their collected blood to the heart.

Another general distinction between arteries and veins is, that arteries alone possess any visible pulsation. The blood in the veins is driven onward by various circumstances extrinsic to themselves; such as the contraction of muscles around them, the pulsation of arteries in their neighborhood, a dependent position, &c. The veins, therefore, have valves, which, when the blood has been squeezed forwards towards the heart, by the adventitious causes just mentioned, prevent its regurgitation, or gravitation backward.

I have said, that there is scarcely any point in the body which is not occupied by vessels and nerves. It follows, therefore, that there is scarcely any point of it which does not in great measure consist of vessels and nerves;—and this is true. When you look at a piece of red raw flesh, that which appears to you a solid mass is, in fact, mainly composed of a wonderful and compact tissue of nerves and hollow tubes, firmly compressed and matted together. The only solid parts are the nervous and muscular threads, a little areolar substance, and the delicate membranes forming the coats of those hollow tubes; that is, the blood-vessels and absorbents;—and even these are porous—at least the blood-vessels.

What I have said of the red raw flesh, is also true of the bones, especially of young animals; for the internal structure of the bones is honeycombed and highly spongoid, and their cells are everywhere filled with vessels and nerves. From all this there results another consequence; which is this—that so large a portion of the body consisting of tubes, and these tubes being filled with fluid, a very large proportion of the whole body must consist of fluid. This, too, is true. If you take a piece of human muscle (that is, what you call, in meat, the lean part) of the size and thickness of an ordinary beef-steak, and dry it perfectly, it will become but little thicker than a sheet of brown paper. In fact, fully nine-tenths of the body are fluid. The next large proportion consists of the solid matter composing the muscles, the nerves, and the coats of vessels. What remains is too trifling for consideration.

Au revoir—adieu!

E. JOHNSON.

LETTER II.

Umberslade Hall, near Birmingham.

My dear John,

In my last Letter, I told you that by far the greater part of the body is composed of a delicate net-work, formed by the interlacing of minute arteries, veins, nerves, and absorbents; and I endeavored to give you a clear notion of the manner in which the arteries and veins are distributed—how they arise, and how they terminate—the differences which distinguish veins from arteries—and also the differences which characterize the two kinds of blood which they contain and convey. I have now to speak of the absorbents and nerves; and explain the manner in which they are distributed throughout the entire body, so as to perform their share in making up that wonderful tissue, of which our organs consist.

The absorbents arise by cul de sacs from numerous points of the body and permeate almost every tissue. They may be compared to a number of long, slender, delicate leeches, attached by their mouth-ends to various points of the interior of the solid tissues; and having their bodies gradually and progressively united, until they all terminate in one tail-end; which tail perforates the side of one of the large veins near the bottom of the neck, on the left side; so that whatever is taken in at the several mouth-ends is all gathered together into the one tail, and emptied by it into that vein, when it becomes mixed at once with the general current of the blood.

A certain and very considerable number of the absorbents arise from the internal surface of the bowels. It is the office of these to take up the nutritious chyle produced by each meal and convey it into the blood.

Now this chyle has somewhat the appearance, and also some of the properties, of milk; and the Latin word for milk is lac—and, therefore, those particular absorbents which arise from the internal surface of the intestines, and which have to perform the duty of absorbing this lac, or chyle, are, on that account, called lacteals.

The fluid in the other absorbents has something the appearance of water; and one of the Latin words for water is lympha—and, therefore, this fluid has received the name of lymph; and on that account, those particular absorbents are called lymphatics, in order to distinguish them from the lacteals. For the sake of perspicuity, I shall call those absorbents which take up the chyle, chylous absorbents, and the others lymphatic absorbents.

Their course is not straight, but waving and devious; and, as they proceed towards their termination, they are perpetually inosculating, that is, uniting, and again separating. They all eventually terminate and empty their contents into the veins.

It was formerly supposed that the office of the lymphatic absorbents is, to take up molecule after molecule of the effete body, convert it into the fluid called lymph, and carry it into the blood; the office of the chylous absorbents being to suck up from the intestines the nutritious chyle, and convey that also into the blood, for the purposes of nutrition. These two sets of vessels, therefore, under that supposition, might be compared, not inaptly, to two parties of laborers;—the one party being occupied in pulling down the old building, and carrying away the rubbish; while the other is equally busy in bringing into the blood new materials, wherewith to rebuild it as fast as it is pulled down; the new materials being carried to every part of the body by the arteries.

This was the doctrine of the Hunters. But modern discoveries, resulting from improvements in the microscope, have now utterly rejected it. It can now be almost demonstrated that the office of the lymphatic absorbents, like that of the chylous absorbents, is also to carry *nutritious* matter into the blood—the effete, that is, dead and useless matter—matters which can no longer serve any useful purpose in the economy—being absorbed by the *veins*, and by them transmitted to the several excreting organs, as the lungs, kidneys, and skin, by which they are separated from the blood and thrown out of the body. As it is the office of the chylous absorbents to carry into the blood certain nutritious matters furnished to them from the solid matters which we put into our stomach as food, so it is the office of the lymphatic absorbents to carry into the blood certain nutritious matters furnished to them from the solid matters of our own bodies. The chylous absorbents feed upon the food which we give them from without. The lymphatic absorbents feed upon ourselves. They are man-eaters.

The food of the chylous absorbents—the food which we eat—is

assimilated to the nature of chyle by virtue of certain changes wrought upon it within the stomach and upper portion of the smaller intestines. The food of the lymphatic absorbents—our own solid flesh—is assimilated to the nature of lymph by virtue of certain changes wrought upon it by and in the little cul de sacs, or cells, from which the mouth-ends of the lymphatics arise. And both the chyle and lymph are further assimilated to the nature of blood by virtue of certain changes wrought upon them in the lacteal and lymphatic tubes and glands.

It is the veins, therefore, which carry off the worn-out particles of the old body; and it is the absorbents (both lacteal and lymphatic), which carry into the blood the nutritious materials, in the shape of chyle and lymph, out of which the daily waste is to be supplied. All these nutritious materials are transmitted through the lungs (mixed with the common blood) to the left side of the heart, by which they are sent through the branching arteries to every point of the body.

As an absorbent passes onward from its origin towards its termination, it every now and then stops, recoils upon itself, and rolls itself up into an irregularly-shaped ball (varying in size from that of a millet-seed to that of a pea), and then proceeds as before. While the absorbent is in the act of forming this ball, it is excessively minute. These balls are exceedingly numerous in the mesentery—that part which, in a lamb, is called "the fry:" they are generally to be found in the neighborhood of large blood-vessels, under the lower jaw, before and behind the ear, at the bendings of the knee and thigh, and in the arm-pit. These little balls are the absorbent glands; and there is scarcely an instance of an absorbent vessel reaching its termination in the veins without having first formed one or more of these glands.

Now, as these glands are merely a congeries of minute absorbent vessels, it is clear that the lymph and the chyle, which these vessels convey, must traverse these glands before they can enter the blood. The chyle and lymph are, in fact, strained through several of these curious little sieves; and this straining produces some necessary alteration in them, by virtue of which they become more nearly assimilated to the nature of blood, as before observed. I need not tell you, after what I have already said of the distribution of arteries and veins, that these latter vessels everywhere accompany, and interweave themselves with, the absorbent vessels and glands.

Now that you understand the nature of the offices and functions performed by the lymphatic and chylous absorbents, and the absorbing functions of the veins, you will easily comprehend what is called the modus vivendi; that is, the manner how we live; viz. in a state of perpetual decay and regeneration—a continual pulling down and building up again. The building up is effected by the minute capillary arteries, which carry and distribute the nutritious materials to every point of the fabric. The pulling down is accomplished partly by the agency of oxygen. The same arteries which carry the nutritious elements, also carry oxygen in combination with these very elements. At the several points at which the body is to be nourished, a portion of the oxygen quits the nutritious elements and combines with a portion of the effete and useless matters, which combination of oxygen and effete matter is immediately absorbed by the veins; and a portion of the nutritious elements which have thus lost their oxygen leaps into the place just vacated by the effete matter. The other part of the pulling down is effected by the lymphatic absorbents, as before observed, and the points thus rendered vacant by the action of these absorbents are refilled by the remaining portion of the nutritious elements which have parted with their oxygen.

These compounds of oxygen with the worn-out body are transmitted by the veins to the excreting organs to be thrown out of the body.

There is not a square inch of you which is the same as it was ten years ago. That which was you, ten years since, is now not you, but something else:—it has been thrown out of the body, resolved into its original elements, has undergone new combinations, and is at this moment, perhaps, flourishing in the shape of some goodly water-dock or field-thistle; or more humbly, but still usefully, employed in stopping the bung-hole of a beer-barrel. "Alexander died; Alexander was buried; Alexander returneth to dust; the dust is earth; of earth we make loam:—and why of that loam, whereto he was converted, might they not stop a beer-barrel?"

The fact is, that we are dying every hour, nay, every instant: and the only difference between this death and the final consummation of life (as far as it regards the body) is, that, in this hourly and gradual death, the place of every dead molecule is instantly supplied by a living one; while, in the other case, all the parts of the body perish at once and together, and are not reproduced.

You are, then, a new and a different being, exercising the same faculties, but doing so with different organs. You still exercise the faculties of vision, of hearing, and tasting; but the eye with which you now see is not the same eye with which you saw ten years ago; it is a new eye; and you hear with a different ear, and taste with another tongue. Indeed, the eye of to-day is not the same as that of yesterday: for a part of the eye of yesterday has been taken up by the absorbents, as food, and converted into lymph; and another part has been conveyed, first into, and then out of the blood, in the shape of perspiration, breath, &c.; while the deficiency thus produced in the eye has been supplied by a part of yesterday's dinner; so that you are now performing the act of vision with a part of the pudding which you ate at that meal. This is not romance, nor speculation, but a literal fact. Is not this, of itself, sufficient to show you the vast importance of the assimilating processes? Does it not clearly demonstrate to you the manner in which faulty assimilation operates, so as to injure the health and perfection of your organs? Is it not manifest that if your assimilating organs do not perfectly assimilate your food, that the deficiencies produced in your eye, by the action of the absorbents and veins, will be either not supplied at all, or supplied with new matter, of an unhealthy quality, so that the new eye will not be so good and perfect as the old one?

From considering the functions which the chylous and lymphatic absorbents perform, together with the absorbing function of the veins and the property of combining with the elements of the worn-out body possessed by oxygen, you will readily understand why we grow in youth, and cease to grow in manhood. It is because, in youth, the building up processes go on more actively than the pulling down processes. In middle life they are in equilibrium. In old age the destructive processes exceed in activity the reparative.

While the manner in which the absorbent glands are formed is fresh in your memory, I may as well describe to you how the secretory glands are formed. This will give you an opportunity of observing the difference between the two; both as it relates to their formation, and to the functions which they severally perform. We have just seen that the office of the lymphatic and chylous glands is to operate some change upon the lymph and the chyle, which has the effect of assimilating them more and more nearly to the nature of blood. But the office of a secretory gland is to elaborate, or manufacture (if I may so speak), out of the blood, a new and distinct fluid; which new fluid is called a secretion; as, the bile, the saliva, &c., &c. Previously to the recent improvements in the construction of microscopes, everything relating to secretion and the secreting apparatus was little else than mere speculation. Now, however, through the agency of these instruments, speculation has yielded to ocular microscopic demonstration.

All the secreting glands are demonstrated by the microscope to consist of a conglomeration of minute cells, everywhere interpermeated by capillary blood-vessels. The secreted fluid escapes from the blood through the coats of these capillaries into these cells, from which it is collected into the excretory ducts which carry it to its place of destination.

This is the way in which all secretory glands are formed. Their size is extremely different, varying from the wonderful minuteness of the ceruminous glands of the ear, whose office is to secrete the wax—and which are, I believe, the smallest glands in the body—to the great magnitude of the liver, which is the largest. But a very large gland is, in fact, only a vast number of very small ones conglomerated into one mass, and united, and as it were glued together, by areolar tissue; and having their several minute excretory ducts united so as to form one, two, and sometimes three larger ones, into which the smaller empty themselves.

I have now to speak of the fourth principal structure which enters into the composition of the ultimate tissue of the body—I mean the nerves. The brain accurately fills the cavity of the skull. With its general appearance you are probably acquainted, from having seen the brains of animals.

The spinal marrow is a tail-like elongation of the brain; which elongation passes out of the head, through a round hole in the back part of the skull. So great is its resemblance to a tail, that it has been called *cauda cerebri;* that is, the tail of the brain.

From the brain and spinal marrow there arise forty-three pairs of nerves; twelve from the brain, and thirty-one from the spinal marrow. The nerves are whitish cords; and every large nerve consists of a bundle of small ones; and these small ones consist of bundles of still smaller, as a skein of thread consists of a number of single threads, and as every single thread consists of a number of still smaller threads, viz., the fibres of the cotton. As a large nerve proceeds from its origin to its termination, every now and then one or more of the threads, of which it is composed, parts company, and takes a course of its own. As these proceed, one or more of the strands, of which they also are composed, disjoins itself from the fellowship of the others, and takes a course of its own: and so on, until the whole have been separated into microscopic filaments of undistinguishable minuteness. You will observe here a remarkable difference in the manner in which nerves are distributed from that in which arteries are given off. The branch of an artery arises from that artery: there is a communication between them; so that the contents of the parent artery flow into the branch which proceeds from it. The large veins, also, are formed by the absolute union of smaller ones; so that the contents of the smaller flow into and mingle with the contents of the larger: but between the large nerves and the branches which proceed from them there is no union nor communication whatever; they are merely in juxtaposition—a bundle of separate threads, bound up together, and inclosed in one common sheath. When, therefore, a nerve gives off a branch, that branch merely parts company, to travel along another road. Every nerve, therefore, however minute, is a distinct thread; having one of its extremities fixed in the brain or spinal marrow, and the other in that point of the body on which it terminates. If it were not for this peculiar arrangement, all our different sensations would be jumbled into one. If we touched a round body with one hand, and a square one with the other, before the two impressions reached the brain they would become mingled; so that the idea which we should derive from these two impressions would be a sort of hybrid idea of a something neither round nor square.

There is one pair of nerves included by me among those arising from the brain, which possesses striking marks of difference from all other nerves: it is called "the great sympathetic pair."—I should have observed, that all the nerves are sent off from the brain and spinal marrow in pairs.—This pair of nerves has given origin to endless discussions: some asserting that it arises from the brain, others, that it does not;—some, that it has one office, some, another. Fyfe says, "It is either formed originally by the reflected branch from the second or the fifth pair; and by one or two, and sometimes three, small filaments, sent down from the sixth pair, whilst in the cavernous sinus: or, according to the opinion of some authors, the sympathetic sends off these small nerves to join the fifth and sixth pairs." Mr. Green says, "This nerve is so essentially distinguished from the other nerves of the body, that it may be described separately, or as a separate system of nerves."

"It consists," he says, "of a considerable number of ganglia (hardish knobs), of which the number and size differ, not only in different individuals, but in the same individual, on the two sides of the body; and of branches, which in part connect these ganglia, or form junctions with the other nerves; and are in part distributed to the internal organs. It extends from the base of the skull, on each side of the vertebral column (backbone), through the neck, chest, and abdomen, as far as the coccyx (that is, the lower extremity of the backbone), forming from above to below, numerous ganglia: those in the neck are few in number; but in the rest of its course it generally forms one ganglion between every two vertebræ (bones of the back): these are severally connected, by one or more filaments, with each other, and with all the nerves of the spinal marrow; and the uppermost cervical ganglion (ganglion of the neck) is connected with most of the cerebral nerves (nerves of the brain). Lastly, it detaches filaments to the viscera (organs of the belly and chest; and those which are distributed to the abdomen (belly) form connections with the numerous set of ganglia in this cavity, which are placed about the trunks of the large vessels."—Thus Mr. Green traces it no higher than the base of the skull; but an anatomist has recently traced it completely round the brain; and thence downward, on either side of the spine, until that portion of the nerve which descends on one side of the spine unites with that descending on the other side, at the extreme lower point of the backbone. During its whole course, there are little knobs situated upon it, at short intervals; so that it has something the appearance of a cord with beads of different sizes strung upon it—or of a chain—or of a small knotted rope, with its two extremities joined so as to form a sort of necklace, as it (the necklace) hangs round a person's neck, falling gradually to a point in front. From the little knobs, or ganglia, numerous nerves are given off, which unite with almost all the nerves coming off from the brain and spinal marrow; and sending numerous filaments also to the organs concerned in nutrition; as the heart, lungs, stomach, bowels, liver, &c., &c. The little knobs with which the sympathetic is studded, have been considered by some as so many little independent brains, whose office it is to supply the organs of nutrition with motive power; and they say that this arrangement was made in order to remove these organs beyond the influence of the will, which has its seat in the brain. The absolute necessity that these organs should not be under the control of the will, and the fact that they are not, together with the additional fact that this pair of nerves does not supply them with motive power, seems, I think, to favor this notion. But, however this may be, it will be sufficiently accurate for our present purpose, to consider all nervous influence as derived from the brain, and from the spinal marrow, which is merely an elongation of the brain.

I have now given you an account of the general structure of the body—sufficiently brief and rough, but nevertheless sufficiently accurate and minute, to enable you to understand the nature of the several functions performed by the several organs of nutrition, whenever I have occasion to speak of these functions and these organs. This general structure is so simple, that you can never forget it. You have only to remember, that whenever you are considering and presenting to your mind's eye any part of the body—whether it be the stomach, the liver, the heart, the bowels, or the arteries and veins—whether it be the solid bones, a mass of flesh an inch thick, or a delicate filmy membrane no thicker than the gilding of your picture frames, it is still the same; it is still nothing more than a matted congeries of arteries, veins, nerves, and absorbents, held together by, and wrapped up in, the meshes of the areolar web. Areolar web is a better term than areolar substance: for, when spread out, it has a good deal the appearance of a spider's web, and has, moreover, of real substance, extremely little indeed.

Take four threads of different colors—a scarlet one, to represent the arteries; a black one, to represent the veins; a white one, for the nerves; and a silver one, for the absorbents. Dip them in melted wax, and then roll them up into a firm ball. This will give you a rude idea of the manner in which minute threadlike vessels can be so arranged as to form a solid mass; for it is easy to fancy three of these threads to be hollow tubes filled with fluid, like arteries, veins, and absorbents. The wax, which everywhere surrounds them and glues them together, will afford you some notion of the principal office of the areolar web; which is to hold the different parts of the intimate structure of the body together, by entangling them in its meshes, as the wax unites the threads by virtue of its stickiness. If, instead of dipping the threads into melted wax, you had dipped them in a solution of phosphate of lime (which constitutes the hard part of bones), the ball, when dry, would have given no bad representation of the structure of the bones.

Now, suppose the former ball—that formed of the threads dipped in wax—to be submitted to a pressure capable of flattening it until it becomes no thicker than a film of tissue paper. This will show you how the same structure which forms the thick, solid, and gross parts of the body, may be so arranged as to form also its most delicate membranes.

A knowledge of the nature and structure of membranes is of the highest possible importance in all that regards the regulation of our diet; for the stomach and bowels are lined with one of those most delicate, and therefore extremely irritable, and highly sensitive, and easily-offended membranes, called the "mucous membrane of the stomach and bowels." It is with this membrane that all which we eat, and all that we drink, comes directly in contact. Here, then, is another powerful reason for caution in what we eat and drink. This membrane is little thicker than gold-leaf; and you know very well that you can scarcely touch a leaf of gold without injuring it—without deranging, and even tearing it. Remember when you are eating your dinner, that the membrane on which every mouthful falls, is little thicker than a leaf of gold.

In contemplating any part of the body—knowing as you now do, that it consists of arteries, veins, nerves, and absorbents—you will please always to bear in mind what are the offices or functions which these structures severally fulfil. You will recollect that it is the function of the lymphatic absorbents to eat away the body—*to feed upon it;* that of the arteries (or rather the vital blood contained in them) to restore what the lymphatics have eaten away; and that of the veins to absorb and carry back to the heart the refuse of the blood; that is, what remains of it, after the arteries have done with it. When the blood has parted with its living elements while in the arteries, the veins carry it away, in order that it may receive a fresh supply of these living elements. But the arteries could not carry the blood from the heart, nor the veins return it to the heart, if they were not supplied with the power of moving. This motive power is afforded them by the nerves—or rather, a fluid conveyed by the nerves. This fluid, however, does not, I conceive, travel along the nerves like a tangible fluid in a tube, but like the electric fluid along a wire. The nervous fluid, therefore, is to the organs of the body what steam is to a steam-engine. And as this fluid is conveyed by single filaments of nerves, it is clear, that wherever there is an artery, vein, or absorbent, there must also be a nerve to enable those vessels to convey their fluids, which they do by a motion of their own or of neighboring parts.

You may conceive the universality of the nerves and blood-vessels by the fact, that you can scarcely insert the point of the finest needle into any part of your body without producing pain and bleeding; which proves that the point of the needle has wounded both a nerve and blood-vessel.

Sis memor mei!

E. JOHNSON.

LETTER III.

Umberslade Park, near Birmingham.

MY DEAR JOHN,

I have already described to you as much of the structure of the body as I believe necessary, in order to enable you to understand the nature of the several actions which are perpetually going on within that structure. It is of these actions that I have now to speak. But previously to a description of the actions peculiar to living beings, it seems proper to devote a few moments to an inquiry into the nature of life itself.

Writers on physiology* are accustomed to enumerate the several distinctive differences which separate the organic from the inorganic kingdom of nature. These are, generally speaking, well marked, and sufficiently understood, by almost every one ; although almost every one might not probably be able to give a scientific

* Physiology is an exceedingly improper term. It is used by the moderns to signify the science of life ;—*animal physiology* being used for the science which treats of the life of animals, and *vegetable physiology* being appropriated to the science of life in vegetables. But the term Physiology no more denotes the science of life, than it does the science of picking pockets. It means the science of nature ; and it is as strictly applicable to the laws which govern inanimate matter, as to those which regulate the actions of living beings. This term, with regard to animal life, should be Zoonomy, which signifies that science which consists in a knowledge of the *laws of life*, and nothing else. With regard to the life of vegetables, the term should be Phytozoonomy, which means the science which makes us acquainted with the laws of *plant-life*, that is, the life of plants. "The endless introduction of new technical terms, on every frivolous pretence," says DR. FLETCHER (a new star in the Iatro-philosophical firmament, and a bright one too), " seems adapted much less to benefit than to injure the cause of Philosophy." True : but when the introductions are *not* endless, and the pretence not frivolous, an exactly opposite result will accrue. Nothing has tended so much to mystify science, and obstruct its progress, as the unsettled state of the exact meanings of words. Words are, through ideas, the signs of things ; and if *one* word be used indiscriminately, as the sign of several things, how is the reader to know which thing of the several the writer desires to indicate ? Dr. Fletcher has himself taken occasion elsewhere to complain, and that loudly, of this improper, indiscriminate use of words.

relation of them. To dwell upon these, therefore, would be foreign to my present purpose. But there are a few characteristics of organic matter, of such vast and immediate importance to all that relates to the preservation of health, that I must not omit to take especial notice of them.

One of the few attributes I shall mention, as peculiar to organized matter, is DEATH.

Death—" the dunnest of all duns,"—death,

> " Sole creditor, whose process doth involve in't
> The luck of finding everybody solvent,"

has been so often personified, indeed, as something horrible—as some " gaunt gourmand," who is by every means to be eschewed —that we are apt to contemplate it as though it were a real entity—a sort of " raw-head-and-bloody-bones," whose chief amusement consists in stopping folks' breath. But I need not tell you that all this is mere rhetorical delusion—one of the poet's " fine phrensies." Death is a sheer abstraction—the mere cessation of life. As the cessation of sound is called silence—as the cessation of motion is called rest—so the cessation of life is called death. Death, therefore, being only the abstraction of life, it is manifest that things which have never lived can never die.

Another condition peculiar and necessary to all matter intended to live is, organism—the consummate result of organization. Organism, in the common sense, is that state of existence in which the elements composing the germs of matter intended to live are held together by a property which may be called " vital affinity," or " the affinity of vitality"—a property which enables it to resist the ordinary agencies of chemical affinities to which common matter is subjected. A seed is an instance in which a germ of matter intended to live (for a seed does not live—it merely possesses vitality, or the aptitude to live) preserves its integrity, in virtue of the vital affinity, and in defiance of the common chemical agencies. A melon seed, a hundred years old, will grow, if planted in a proper soil.

But the term organism is not only used to indicate a peculiar condition of the elements of matter, but also a peculiar condition of masses of matter. Here it signifies that state of existence in which masses of matter grow and preserve their integrity by virtue of a power which may be said to consist in the affinities of assimilation—a power withdrawing them from the influence of common chemical agencies, until they shall have accomplished the final cause of their organization—a power enabling them to assimilate other matter to their own nature and substance.*

Another most important characteristic of living matter is its contractility; that is, not contraction, but the power of contracting: it is the being able to contract.

Now, Indian-rubber, or a steel spring, may be said to be able to contract. But then the one of these can only do it after having been put upon the stretch, and the other only after having been bent; they can only contract after having been put into an unnatural condition. In their natural condition they are, like all other inorganic matter, at rest; and can neither contract, nor expand, nor dilate, without being first submitted to the action of mechanical or chemical force. These, therefore, are merely elastic. But living matter can do much more than this. When at perfect rest, and in its natural state, it can contract, shrink, and, in short, perform spontaneous movements, merely on being excited, stimulated, or irritated, without the agency of any mechanical or chemical power. It does this by virtue of a property called contractility. When you look at a very strong light, the iris, the colored part of the eye, being irritated by the rays of this strong light, contracts, and almost closes the pupil; that is, the black spot in the eye, which is, in fact, a round hole. When your will directs your arm to move, the muscles of the arm, stimulated (that is, excited) by your will, contract, and raise the arm accordingly. When the blood rushes into the right side of your heart, that part of the heart contracts, and pushes it into the left side: then the left side contracts, and pushes it into the aorta. All these contractions could not of course be executed, if it were not for the property of contractility—that is, the ability to contract.

Now, all the motions of the different parts of the body, without and within, are performed by these contractions, and by virtue of this contractility. It is the main-spring of the watch—it is the

* It is perfectly correct to call the assimilating processes by the term "affinities of assimilation." For what is affinity, but an alliance or relation? And is there not a relation between the food and the body which it nourishes? Is there not an alliance between them? And what are the assimilating or nutritive processes or actions, but those actions or changes wrought on the food, by which its alliance to the body is drawn closer and closer, until they become identical? And so it is correct to say, that the proximate atoms of organic germs are *held together* by *vital* affinity; for this is *not* equivalent to saying they are *brought* together by vital affinity.

chief wheel in the machine—it is the principal beam, the main prop, of the building: by it we gather our food—by it we eat it—by it we swallow it—by it the stomach sends it on to the bowels: from the bowels it is carried to the heart by it; and by it having become blood, it is circulated through the body, for whose nourishment it is destined. Whenever, therefore, I use the term contractility, you will know that I mean that power by virtue of which the several parts of the body are able to move, and perform those actions which are proper to them.

A third property, distinguishing organized from inorganized matter, is sensibility.

This is exceedingly slippery ground, and rendered still more dangerous by the darkness in which it is enveloped. I shall therefore hasten off the ice as quickly as possible, lest some invisible straw or other should trip up my heels.

"Irritatio," says Glisson, "est perceptio, sed sensatio est perceptio perceptionis;" that is, "Irritation is perception, but sensation is the perception of a perception." Said I not it was slippery ground? But Dr. Fletcher, speaking of this definition of Glisson, says, "for either terseness or accuracy, it cannot, perhaps, be improved." To me, however, I confess, it has very much the appearance, not of splitting a hair—that is but a trifle—but of splitting the very ghost of a hair, which is no trifle. Lobstein defines sensibility as "facultatem stimulum percipiendi;" that is, "the faculty of perceiving a stimulus." You probably know that anything which irritates or excites any part of the body to action is called a stimulus. I think Lobstein is right. Thus the heart, by virtue of its contractility, has the power of contracting; but it is by virtue of its sensibility that it perceives the proper moment for exerting this power—the precise time when to contract; viz. when the blood stimulates it by its presence, as it rushes into its cavities. Sensibility, therefore, is that property of organized matter by which it becomes aware of an impressing cause—by which it perceives when it is acted upon by a stimulus.

In vain would the heart be organized—in vain would it be endowed with contractility (that is, the power of acting)—were it not also endowed with sensibility; that is, the power of knowing when to act—of feeling the presence of a stimulus. The several stimuli may be likened to a number of messengers sent out from "head-quarters," the heart, in order to tell the several parts of the body when to act; and the arteries are the roads along which they travel

—the principal stimuli within the body being the countless streams of blood continually flowing through its arteries. You must remember, however, that this office of stimulation is by no means the main duty which the blood has to perform: it is only an adventitious office—only one of the numerous functions which the blood performs. Besides the blood, there is another remarkable stimulus—another messenger sent to certain parts of the body, to summon them to action; which is sent, not from the heart, but from the brain. This messenger is a strange, incomprehensible being; and his name is *Will.*

Comparing organized matter to a musical instrument, and its aptitude to act, i. e. live, to that instrument's aptitude to sound, one might liken the stimulus offered by the blood to the performer, whose office it is to play upon that instrument.

These two properties, sensibility and contractility, constitute vitality. I say vitality—not life.

And here allow me to caution you against the common error of confounding vitality with life. The term "vitality" no more signifies life, than the word fiddle signifies music. Vitality signifies, not life, but livability (if I may coin a word); that is, the aptitude or fitness to live; as musicality (if I may be allowed to coin another word) would denote, not music, but the aptitude or fitness to give rise to musical sounds. Vitality is a secondary cause—a necessary condition of organized matter, in order to give rise to living actions; as musicality would represent a necessary condition in a fiddle, in order to give rise to musical sounds. A fiddle may be perfect in all its parts, and yet, for want of this necessary condition, which I have called musicality, be wholly unable to produce musical sounds. For instance:—if you were to fill the body of Paganini's best fiddle with sand, and soak its strings in tallow, Paganini might go mad, perhaps; but twenty Paganinis, or one Paganini with a twenty-Paganini power, which is the same thing, would not be able to extract from it a single musical tone. Why? Because the instrument would have lost that necessary condition which I call musicality—the sand and the tallow have destroyed it. "En caput; sed cerebrum non habet;"—which, being interpreted into the vulgar tongue, for the benefit of "ears polite," signifieth, "There is the fiddle; but where is its aptitude to discourse most excellent music?" I will make this clear in a moment. The first condition necessary to life is organism—that is, the fiddle; the second is vitality, or that condition or manner of existence necessary to the production of living actions—that is, the musicality, or that particular mode of a fiddle's existence necessary to the production of musical sounds ; viz., perfect freedom from sand and tallow, and all other impediments to musical sounds. And, as we have just seen that a fiddle may exist perfect in all its parts, and yet be wholly destitute of musicality, and therefore entirely unable to emit sound ; so organized matter may exist, and yet, for want of vitality, be wholly unable to live. I know a man who has a very large wen, situated on the back of his head. If this wen were shaven off, it would still for a time remain perfectly organized—but it could no longer live. Why ? Because it would have lost its vitality—that condition necessary to life, which, in this instance, would be an endowment derived from its connection with the rest of the man's body.

And again: as organism may exist without vitality, so may vitality without life. Seeds are an example of this. A grain of mustard-seed does not live: in it there is neither motion nor fluid; and it is utterly impossible for a moment to conceive the existence of life without both these. But it possesses the aptitude—the ability to live—that is, vitality ; and if you plant it in a proper soil, it actually will live, and become possessed both of fluid and motion. A grain of sand, on the contrary, possessing neither organism nor vitality, will remain a grain of sand forever—plant it in whatever soil you please: at least, it cannot undergo any changes but such as are purely chemical or mechanical.

As vitality is not life, then, so neither is it organism, but merely a condition of the latter, necessary to the existence of the former. Life, then, being neither organism nor vitality, what is it ?

"Life," says Richerand, "consists in the aggregate of those phenomena which manifest themselves in succession for a limited time in organized beings."

"Life," says Dr. Fletcher, in one of the most erudite, elegant, and ingenious works that ever fell from the press—"life consists in the sum of the characteristic actions of organized beings, performed in virtue of a specific susceptibility (vitality), acted upon by specific stimuli." These two definitions are perfectly consentaneous with each other; and to them I have nothing to add. Life, like death, is not an entity: it is merely an aggregation of effects. To say what life is, is only to enumerate all the actions of which a living being is capable; not only the visible actions, as of the members, but also the molecular actions, as those invisible motions among the ultimate and proximate molecules of the matter of which he is composed, and by which his nutrition is effected. Life is, to organism, contractility, sensibility, and stimuli, what chemical phenomena are to chemical affinity—what astronomical phenomena are to the centrifugal and centripetal forces, and the antagonization of these forces by each other—what the motion of the hands of a watch is to the mainspring and its elasticity; viz., the sum total of numerous effects, of which these four attributes of organic matter above-mentioned are the secondary causes. These effects we call living actions—actions, the totality of which constitutes life.

Organized matter is a harp, of which vitality is the musical power: stimuli are the fingers of the performer: and life is the music produced—a hymn, day and night, in praise of the goodness and power of Him, who permits

> "This harp of a thousand strings
> *To keep in tune so long.*"

Such is life.—Now, what is health?

As life consists in the aggregate union of all the living actions, and indifferently, whether those actions be well or ill performed; so health consists in the aggregate union of such only of those actions by which nutrition is carried on—and not indifferently, whether they be well or ill performed; but exclusively when they are well performed. And disease consists exclusively in their being (one or more of them) ill performed, or not performed at all.

You will now readily understand of what tremendous importance to health are the properties contractility and sensibility; for, as health consists in the due performance of certain actions, and as these actions depend on contractility and sensibility, it is clear that they will be feebly or energetically performed, accordingly as these two properties are themselves in a healthy condition. You will also see, that the stimulus which the blood offers to them is of vast importance likewise. This stimulus is a sort of messenger, sent to summon them to action. In proportion as the summons is feebly delivered, it will be faintly heard, and feebly obeyed. Contractility and sensibility are like a horse that gallops furiously, moves sluggishly, or goes to sleep entirely, exactly in proportion as the stimulus of the whip is gently or vigorously applied. Like the horse, too, the faster they are urged forward by the whip, the sooner they become tired ;—like him, they may be driven even to death ;—like him, they require rest and repose. Do not therefore be led to undervalue the importance of these properties, because of the playfulness with which I have occasionally spoken of them ; as, for instance, in the allusions to Paganini and his fiddle. "Ridentem dicere verum quid vetat?" What reason on earth is there for always telling the truth with a grave face ? Why should we not sometimes tell it with a smiling eye, as well as a scowling brow ? Gravity is not wisdom ; nor a smile, folly. Besides, if to smile be a folly, what then ? "Qui vit sans folie, n'est pas si sage qu'il croit," says Rochefoucauld. He who lives without folly is not so wise as he believes himself.

Organism, then, is that arrangement of the component parts of matter which fits it to be endowed with contractility and sensibility. Contractility is that property which endows it with the power of executing living motions. Stimuli are impressing causes, acting on its contractility, and exciting it (organized matter) to action ; and sensibility is the property by which it perceives the presence of these impressing causes.

The muscles of your arm are organized, and they possess contractility and sensibility : and when you resolve to raise your arm, your will becomes an impressing cause, exciting those muscles to action ; that is, to contract. Their sensibility makes them aware that this impressing cause, or stimulus (viz., the will), is acting upon them ; and they contract in obedience to it ; and your arm is raised accordingly.

I trust, my dear John, there are now clearly depicted on the canvass of your mind, four distinct and well-defined ideas, representing organism, contractility, sensibility, and stimuli ; and that you plainly perceive their intimate connection with each other, and the necessary co-operation of all, in order to produce the phenomena of life. As to stimuli, when you consider the literal meaning of the word, you will have no difficulty in understanding that modified sense in which it is used in the language of science. It means, literally, a long stick, with a sharp point, with which husbandmen were wont to goad their oxen along, in times and countries when and where oxen were used for agricultural purposes.

Now these properties, contractility and sensibility—these important properties, upon which, it is manifest, life depends ; and without a healthy condition of which the health of the body can no more be preserved than the true motions of a watch can be maintained with a broken or otherwise injured main-spring—these properties, I say, are subject to certain laws. I am now about endeavoring to establish these laws; or rather, I am going to endeavor to convince you of their existence. That they do exist, is a truth that has been well and incontrovertibly established and admitted among all men acquainted with the animal economy, the effect of medicines upon it, &c., &c., ever since Hippocrates practised physic at Athens; and that is more than two thousand years ago. But it is not sufficient that medical men are aware that these laws exist;—my object is, to convince you of their existence. I want you to know what is good and what injurious to your health; not from my dictum, but from the simple exercise of your own reason.

I beg that you will consider what I am about to say on the subject of these laws with great attention;—examine the proofs and arguments carefully, but fairly. For I tell you, at the outset, that if you admit the existence of these laws, you will not afterwards be at liberty to question or doubt the truth or propriety of what I shall say with regard to diet and regimen. For the existence of contractility and sensibility are like the axioms of Euclid;—they are self-evident truths, of which any one may convince himself by experiment. For instance, a dead man may easily be made to move his limbs, to breathe, to frown, &c., by exciting the appropriate muscles to contract by means of galvanism. And the laws to which these properties are subject, and of which I am now to speak, are, if I prove them, of the nature of the propositions of the First Book of Euclid. If these be true, the propositions of the Second Book must be true also, of necessity—the truths of the Second Book arising out of the truths of the First. As, for instance, if you admit that twice two are four, you must of necessity also admit that the half of four is two. So if you admit what I am about to say of these laws, you must also admit the propriety of what I shall hereafter say as to diet and regimen; as the correctness of the latter will depend solely upon the correctness of the former. When I come to apply these laws to the subjects of diet and regimen, I repeat, that either what I shall say then must be true, or what I am about to say now must be false.

As all the actions of the body are performed by contractions, and as these contractions are performed in virtue of the contractile power, that is, contractility, it is evident that the physical strength of the body—that strength by which we raise heavy weights, walk, run, leap, &c.—will be in proportion to the energy of the contractile power. A high degree of contractile power, then, is synonymous with strength; and a low degree of contractile power is synonymous with weakness.

But not only are the motions of the limbs performed by contractions, but also those motions of the internal organs by which nutrition is effected. Now, this being the case, and as these internal contractions are also performed in virtue of the contractile power, or contractility, it is again manifest that the energy with which these internal motions are performed (and by which nutrition is effected) will be also in proportion to the energy of the contractile power: and as health consists in the due energy (as we have before seen) with which these motions or actions are effected, it follows, clearly and logically, that a high degree of contractile power is synonymous with a high degree of health; and that a low degree of contractile power is synonymous with feeble health.

Having premised the above short paragraph, I now proceed to mention to you the first important law to which contractility is subject, viz. *evanescence*. Contractility can only exist in perfection in recently-organized matter. No sooner has a molecule of matter become organized and assimilated to the living matter, than its contractility begins to fade—to evaporate, as it were, like breath which has been breathed upon a highly-polished surface, such as steel, or looking-glass. Indeed, it seems to be the evanescent nature of contractility which has given occasion to that particular contrivance by which life is supported; viz. by constant organization and disorganization; that is, perpetual building up and pulling down. For if contractility could continue to exist in full energy, in an organized body, during the whole time that body was destined to exist, what necessity was there for this constant renewal—this constant disorganization and reorganization—this constant pulling down and building up?

The evanescent nature of contractility may, I think, be accounted for thus:—It seems to have been a predetermined law of nature, that the only permanent condition of matter should be the inorganic condition. Nevertheless, certain ends in the general scheme of creation were to be fulfilled which required for their accomplishment the existence of organized matter. But, in order that organized matter might not be permanent, and so destroy or neutralize that original law by which it was enacted that there should be no permanent condition of matter except the inorganic, all organized matter was made subject to the laws of fermentation and putrefaction, as they are usually called; whose office is to destroy its organism, and bring it back to its original inorganic condition. But if this had been all that was done, the objects for which matter had been organized could never have been accomplished; for no sooner would matter have become organic, than it would instantly have begun to be disorganized again, by virtue of the laws of fermentation and putrefaction, to which it has been made subservient. But the ends to be answered by organized beings required time—required a continuity of existence, in a perfect state of organism, for a determinate period. It was necessary, therefore, that there should be another contrivance, in order to withdraw organized beings beyond the influence of the laws of putrefaction and fermentation for a definite time; that is, until the purposes for which it had been organized should be accomplished. The phenomena of life exhibit this contrivance—a number of temporary phenomena, set up in order to withstand the phenomena of fermentation and putrefaction for a limited period. But, since the phenomena of life resulted from contractility, and because contractility can only reside in full activity in very recently organized matter, it was necessary, in order to make the matter of an organized being a fit residence for vigorous contractility, that it should be continually renewed; that, while the whole being, as a being, grew older and older, the molecules of which he is composed should, nevertheless, be always young. And thus we observe in the aged, in whom the process of renewal goes on but feebly, and in whom the laws of fermentation and putrefaction are gradually gaining the ascendency over the laws of life—the laws of that contrivance which was instituted in order to remove living beings, for a time, from the influence of fermentation and putrefaction—we observe, I say, in the aged, that contractility is greatly diminished—it has waned, it has faded—their strength is greatly reduced—they are no longer a fit residence for active contractility; since this property can only reside, in its perfection, in very recently-organized matter; whereas, in the old, organization goes on very slowly and imperfectly. On the contrary, in children, contractility exists in a very high degree; because, in them, the process of organization goes on with great rapidity. A child will romp about on its legs for a whole day, without feeling fatigue;

and can endure exertion far longer than a man, when we take into consideration the comparatively small size of the child's muscles.

It is a law, then, of contractility, that in order to its perfection, it is necessary that the molecules of the parts in which it resides should be rapidly reorganized—in a word that they should always have a plentiful supply of healthy and well-vivified blood; for it is out of the blood that the solid body is repaired—reproduced.

Another proof of the evanescence of contractility is, the physical weakness which invariably attends total inaction.

Another law of contractility is this: It is in perpetual strife with the laws of fermentation and putrefaction. This law arises necessarily out of what I have just said; viz., that life is a contrivance to withdraw, for a time, organized beings from the influence of fermentation and putrefaction. It is proved also by the fact, that healthy living beings cannot putrefy—that the universal solvent, the gastric juice itself, has no power to dissolve living animals when accidentally taken into the stomach—that living beings, in whom life and strength (that is, contractility) are at a very low degree indeed, as in putrid fevers, do begin to putrefy partially—and that all beings who have lived, are instantly acted upon by the fermentative and putrefactive forces, as soon as contractility has left them: observe, *not* as soon as life has left them; for contractility will sometimes remain for a short time after life has ceased. Contractility, you must remember, is not life, but one of the secondary causes from which life results.

Another law of contractility is, that it is in an inverse ratio of sensibility. When contractility is vigorous, sensibility is dull; and when contractility is deficient, sensibility is acute. This will be proved, when speaking of the laws of sensibility.

In my next Letter, I shall speak of certain laws and characteristic facts peculiar to sensibility. Till then, adieu!

E. Johnson.

LETTER IV.

Umberslade Hall, near Birmingham.

My dear John,

I am now to speak to you of certain laws, or circumstances, by which the sensibility of the body is materially influenced.

I have elsewhere noticed, that it is by means of the organs of our senses that a proper relation is established between ourselves and the various natural objects with which we are surrounded. It is by means of these that we are able to appreciate the value of these objects, and their power of affecting us, whether injuriously or beneficially. It is by these organs that we are able to discover the means of avoiding whatever is hurtful, and of selecting and securing whatever is necessary to our comfort and well-being. The eye warns us of the approach of danger from before; the ear, from behind; while the senses of smell, taste, and touch, enable us to decide upon the qualities of whatever matters are presented to us for food. But the medium through which these organs are enabled to render us these important offices, is their sensibility. For if the eye were insensible to light, we could not see—the ear to sound, we could not hear—the nose to odors, we could not smell—the tongue to flavors, we could not taste—the skin to touch, we could not feel. In literal fact, then, you see it is sensibility, after all, which establishes this necessary relation, of which I have spoken, between ourselves and surrounding objects; the organs being no more than the instruments by which sensibility exerts its influence. Sensibility, then, is our guardian angel;—it is, like the sailor's "sweet little cherub," an invisible agent, that forever watches over "our lives and safeties all."

Every organ has a kind of peculiar sensibility of its own. Thus, the sensibility of the eye is not affected by the stimulus of sound; nor can the sensibility of the ear take cognizance of the stimulus of light. The nose is insensible to the stimulus of flavors, and the tongue knows nothing of odors. From this it follows, that the sensibility of each organ is adapted to be properly affected by certain stimuli only. All others than these will either not affect it at all, or affect it painfully and injuriously. Thus sound, being a stimulus proper to the ear, but improper to the eye, will affect the ear properly, but the eye not at all. Again, salt is a stimulus proper to the stomach, and, when it comes in contact with the membrane which lines that organ, it affects its sensibility agreeably and healthily; but if you blow salt into your eye, it will produce the most violent pain; yet the membrane which lines the stomach is as delicate in its texture as that which covers the eye. Thus, again, there are certain medicines which exert their influence only on certain organs. Some will act on the stomach, some on the bowels only; some on the kidneys, some on the brain, some on the liver. If you rub belladonna into the skin of your leg, it will not affect your leg; but you will be astonished to discover that you have suddenly become blind. This once occurred to a patient who was under the care of the late Mr. Abernethy for a sore leg; Mr. Abernethy having ordered the sore to be dressed with the extract of belladonna. The man, however, recovered his sight. Mr. Abernethy never dressed sore legs with belladonna again. I heard him relate this circumstance myself.

Every organ, therefore, has a peculiar sensibility of its own, and can be properly affected by certain stimuli only—all others, if they affect it at all, affecting it injuriously; and the evidence of the impropriety of a stimulus is the pain, or some other inconvenience, produced. Hence arises a corollary; viz., that whatever stimulus produces pain, or other inconvenience, is an improper stimulus. The pain, for instance, produced by blowing salt into the eye is sufficient proof that salt is a stimulus not proper to that organ, and cannot, therefore, be applied to it without injury.

This peculiar, distinctive, or eclectic, sensibility is impaired by over-stimulation. Thus we may be deafened by excess of sound, and blinded by excess of light. Everybody knows, too, that snuff will produce, in persons not accustomed to it, violent and painful sneezing; while those who have been accustomed to it for some time can take the strongest kinds with apparent impunity. Again, persons who have never smoked tobacco will generally be sick when they first begin to do so; but, after a short time, they can smoke pipe after pipe without apparent inconvenience. If a person not accustomed to drink anything stronger than water were to swallow a glass of whisky, it would almost choke him; while a Scottish

Highlander will toss off glass after glass, not only without inconvenience, but with a most pleasant gusto.

Now, what have these persons done—these pipe-smokers, and dram-drinkers? Why, as far as the organs in question are concerned, they have, by blunting their sensibility, actually thrown dust into the eyes, and partially blinded that very "cherub," whose sole business it is to watch over their safety.

When a man, for the first time, swallows a glass of raw spirit, his guardian angel, sensibility, tells him—not indeed in a language that can be heard, but in one far more impressive—a language that can be felt—tells him, I say, as plainly as pain can speak, that raw spirit is an injurious stimulant. Yet what does he do? Why, turns a deaf ear to the intimation which could be, by possibility, no other than a friendly one, and obstinately perseveres till the voice that warned him warns him no more;—and then, with a folly scarcely less than idiotic, and an impudent hardihood scarcely less than blasphemous, he exclaims:—"Behold! it does me no harm! it gives me no pain! it causes me no inconvenience!" thus appealing, in his defence, to the silence of that voice which he has himself forcibly silenced. This is abominable!—Let every man drink what poison he pleases;—of this I do not complain:—but let him not attempt to defend the practice: for this is to allure others into the same trap which is already closing its iron teeth upon his own hapless person.

But, luckily for us frail mortals, when this natural sensibility has been only impaired—not utterly destroyed—it can be restored by rest, and only by rest; that is, by ceasing to stimulate it. A few common and well-known facts will be sufficient to prove this. If a man has taken snuff for ten years, and then left it off for ten years, should he begin again, he will be as much affected by it as he was at the first pinch. If a man spend an hour in the belfry of a church while the bells are ringing, when he comes down he will be almost deaf for a time: shortly, however, he will recover his hearing. If a man look at the sun for a minute or two, when he first looks aside he will not be able to distinguish objects: he will, however, presently recover accurate vision. If a man has drunk spirit till the lining of his throat has no more sensibility than the lining of a copper-kettle, let him desist for a few months, and it will be restored. If a man have fed on highly-seasoned soups, piquant ragouts, and other French abominations, until he can discover no flavor in dry bread, let him be sent to Brixton tread-mill for a month, and he will discover that a penny loaf is a delicious morsel. But I need not multiply instances;—your own recollection will furnish you with abundant proofs, that the way to restore impaired sensibility is to allow it to remain for a time unstimulated.

Another peculiarity of general sensibility is, that it can be accumulated in one organ—drawn from all other parts of the body, and concentered, as it were, in one. The insensibility to pain (I mean of course comparatively) which madmen possess, is well known; and several remarkable proofs of this are given by Dr. Hibbert, in his "Philosophy of Apparitions"—a book which you, and every one else, ought to read. We know, too, that persons under the influence of engrossing attention may be spoken to, and even plucked by the skirt—" plucked by the ear," as Horace says—without their perceiving it. There are irresistible proofs of this to be drawn from natural history, but it would be improper to mention them in a work like this. We know, too, that when any one part of the body is in great pain, the rest of the body is nearly insensible to lesser pain. This fact has given rise in the Tonga islands to a curious operation for the cure of traumatic locked-jaw. It consists in inflicting on the wretched patient, in some part of his body, a pain, the anguish of which shall be greater and more excruciating than the tetanic agony: thus, as it were, restoring the equilibrium of the sensibility; and subduing a great pain, by inflicting, for the time, a much greater. The operation is said to have been successful: but the operators complain, that they can get few patients to submit to it. Sensibility, then, can be drawn from one part of the body, and concentrated in another.

Another curious circumstance connected with *sensibility* is *sympathy*. All the organs of the body appear to sympathize with one another. That the brain and stomach have a strong sympathy with each other is absolutely certain: the proof of which is, that a violent blow on the head will produce vomiting; and a deranged state of the stomach will produce headache. Whatever, therefore, affects your stomach injuriously, will also affect your brain injuriously. He, therefore, who ill-treats his stomach, does so, not only at the expense of his bodily health, but at the expense of his understanding also; that is, at the expense of his mind's health.

From this law of sympathy, then, it follows, that you cannot injure any one organ of the body without also injuring, to a certain extent, the whole of the body, and the mind also. And, indeed, when we consider the intimate connection which exists between all the organs of the body, and the reciprocal dependence of all upon each other—when we remember that all the nerves have one of their extremities fixed in one or other of the great nervous centres and their other extremities distributed to every point of the body—when we recollect that all the organs are made up, not of separate net-works, but of different portions of one and the same universal tissue—when we consider this, there seems nothing surprising in the existence of this sympathy. The body, although composed of numerous parts, forms, nevertheless, one harmonious whole; and you cannot remove one part without injury to the whole. And again, although each organ performs a distinct function or office, yet it cannot do this without the co-operation of others. Thus, the stomach can exert no influence on the food, unless it be well supplied with blood by the heart. If the brain die, the heart must cease to pulsate; and if the heart cease to move, the brain must necessarily perish. It is not, then, by any means surprising that an injury to one organ should be felt by another;—on the contrary, it would be very wonderful if it were not so. Accordingly, we see that a wound of the foot will often produce locked-jaw;—the disease, in this instance, being at one extremity of the body, while the wound that produced it was inflicted on the other. If it were necessary to make this mutual dependence of all the organs on each other still clearer, we have only to recollect, that it is impossible to injure any one wheel of a watch without injury to the whole machine—without incapacitating it properly to fulfil the office for which it was intended. And man, the master-miracle of nature, is a machine of far more delicate and complicated structure than a watch, and therefore more readily deranged.

Now, if no organ concerned in the preservation of health (for it is of these I am speaking—the organs of nutrition) can be disordered without disordering all the others, how much more certainly (if this were possible) will it be the case, when the stomach, one of the most important of these organs, as well as the first which is called into action in the process of nutrition, is kept in a state of almost perpetual excitement and unnatural activity. How can it be conceived to be possible that the other wheels of the machine can go aright, when this, the very first wheel which is put in motion, and on which the motions of all the others depend, goes wrong?

Let this, then, be engraven an inch deep on the tablet of your memory—that you cannot injure *one* of the organs of nutrition, without injury to the whole.

I now come to the last law of sensibility which I shall mention; and it is, where all are important, the most important. Therefore, my dear John, draw your chair, put your feet upon the fender, snuff the candles, adjust your spectacles, and read with attention: for I deny it—no man can read attentively who has not first fixed himself in a comfortable position.

This is the law; viz., that sensibility is always in an inverse ratio of contractility. When contractility is vigorous, sensibility is low; and when contractility is feeble, sensibility is always acute; and as vigorous contractility is synonymous with health and strength, the greater or less degree of sensibility becomes an infallible test of the state of the health.

I have already proved that it is the sensibility of our organs which establishes the necessary relation between ourselves and the objects which surround us. From this it follows directly, that it is upon sensibility that all our pleasures and all our pains depend: for there is no pleasure and no pain which is not derived to us from impressions made by external objects upon our feeling—of which feeling, sensibility is the soul. I mean the feeling as well of the mind as the body.

Now, the sensibility of a perfectly healthy man, is so regulated, as to afford him the greatest possible degree of pleasure with the least possible amount of pain; that is, consistently with his terrestrial existence. Indeed, our pleasures are the voluntary and bountiful gift of nature. For our pains, we have nobody to thank but our foolish selves. So good has the great Governor of the universe been to us, that we could not, if we would, escape pleasure; but in almost every instance, we can avoid pain, if we will: for pain is only a warning voice, intimating to us that we have got into a false position—that we are doing something which we ought not to do, or leaving something undone which we ought to do;—in a word, that the proper relation between ourselves and surrounding objects has been, for the time, destroyed. Man, if he would be content to be what nature made him, need scarcely know what pain is.

But we have quitted the position which she appointed for us—we have forsaken the habits which she allotted to us—we have disregarded her tokens—derided her counsels—broken her laws—overleaped her boundaries—and now that we are paying the penalty for our frolic, we stand gaping at each other like fools, and wonder what is the matter! Why, we are like Rabelais' wooden legs—we are like square men that have got thrust into round holes! No wonder we are uneasy—no wonder we suffer pain—we do n't *fit* our position.

Nature, then, has endowed us with a certain degree of sensibility; and my object is, to show that we cannot increase this without diminishing our proper amount of pleasure, and augmenting our proper share of pain—without enfeebling our contractility; in a word, without injury to our health and strength.

Pain is invariably the result of impressions too acutely or strongly made: and, as sensibility is that property by which impressions are felt, it is perfectly manifest that the more acute sensibility is, the more acutely impressions will be felt. And thus those impressions, which in a healthy state of the sensibility afford only pleasure, become painful; and those which always produce pain become more painful. This is so very clear as to render amplification wholly unnecessary.

In order to prove, that wherever there is a high degree of sensibility, there is always a low degree of contractility, i. e. strength—and that, wherever there is a low degree of strength, there is always a high degree of sensibility—you need only look through the world.

Let us first approach the couch of sickness. Tread lightly;—for the slightest noise makes the poor sufferer start, and gives him the headache. Be careful to close the door after you; for the faintest breath of air gives him cold. See how he is shading his eyes with his hand! for the few rays of light which struggle feebly through the venetian blind are painful to him. Observe his hand: how white and bloodless! If you take it in your own, you must handle it as you would an infant's—an ordinary pressure will make him cringe with pain. His banker has just failed, and reduced him to ruin; but you must not breathe a syllable of this in his hearing!—it would kill him. Do you observe that rope suspended over the bed from the ceiling, with a small cross-bar of wood attached to the end of it? So faint is the contractility of his muscles, that he could not, without this contrivance, raise himself in bed. Observe him, as he carries his cup of gruel to his pallid lips! Mark how the liquid quivers in the vessel! Hark, how its edge rattles against his teeth, as he applies it to his mouth! The contractile property of the muscles of his arm is so feeble, that they have not power to keep the limb steady, even while he carries nourishment to his mouth. His heart, too, contracts so feebly, that it cannot send the blood far enough to reach the skin. It is this that makes it so deadly pale;—it is this, too, which make him shiver on the application of the slightest current of air.

In the above picture you will observe two things: first, that the contractility of the invalid has almost entirely disappeared, leaving him powerless; and, secondly, that his sensibility is so acute, that those impressions of light, sound, touch, &c., which under ordinary circumstances were only necessary to the enjoyment of existence, have now become sources of painful suffering; thus proving that whenever sensibility is advanced beyond the natural standard, the sources of pain are multiplied, and those of pleasure diminished; and that wherever sensibility is excessively high, contractility (that is, strength) is excessively low.

It is true, that this is a case of extreme illness, and that every departure from health will not produce this extreme state of things. But it will produce an approximation to it. In the slightest departure from health, the same effects will be produced; the only difference being in degree.

To prove this—let us take a peep into "my lady's chamber." Here you will find the same circumstances of heightened sensibility, and depreciated contractility, which you observed in the sick room—only in a less degree. It is true, that she can bear an ordinary degree of light without pain, and that the sound of your footfall may not give her the headache; but if you leave the door ajar, she will most likely take cold;—if the force of your friendship cause you to press her hand a little too forcefully, she will assuredly scream;—and if you steal slyly behind her when she thinks she is alone, and cry, "Bo to a goose!" she will in all probability fall into hysterics. If you press her arm strongly between your finger and thumb, you will make it black and blue;—while it would require, in order to produce the same effect on one of Mr. Barclay's draymen, little less than the gripe of a blacksmith's vice. "The hand of little employment hath the daintier sense."

So much for her sensibility;—now for her contractility. Could she carry a bushel-basket of potatoes upon her head, for a mile, without resting? Not she. Yet why can she not? It is true, she is a lady; but, as Burns says,

> "A man 's a man for a' that."

And is not a woman a woman for a' that? There is many a woman, neither so tall nor so well-proportioned, who would carry a bushel of potatoes on her head, without resting, from Penny Fields to Pedlar's Acre, and think herself well rewarded with a shilling. There must be some better reason for this great difference, than the mere fact of one being a lady, and the other a woman. The true reason is, that the contractility of my lady's organs, especially her muscles, has sunk exactly as much below the natural standard as their sensibility has been elevated above it.

Thus, then, we have indisputable proof that sensibility and contractility are always in an inverse ratio of each other. But you must be careful to observe, that it is not increased sensibility that gives rise to diminished contractility; but it is diminished contractility that increases sensibility;—increased sensibility being no more than the proof of diminished contractility; that is, enfeebled health and strength.

From this it follows, therefore, that the degree of sensibility depends upon the degree of contractility; and this is fortunate, for we have on that very account only to raise contractility, in order to lower an irritable, acute, troublesome, and unhealthy degree of sensibility to the natural standard. This can be easily done; so easily, that I will undertake, within one month, without fee or reward, or pill or potion of any kind, and on the peril of my head, to enable any lady within the pale of the United Kingdom of Great Britain and Ireland, to carry a bushel of potatoes on her head from Penny Fields to Pedlar's Acre without resting, and that with no more pain and labor to herself than may be sufficient to spare her pocket the expense of sixpenny-worth of rouge.

Now, then, it is quite clear, that whatsoever causes, circumstances, regimen, or conduct, have a tendency to heighten sensibility, must necessarily have a tendency to depress contractility; since I have proved that a high degree of the former is wholly incompatible with a high degree of the latter, and therefore cannot exist in conjunction with it.

The reason why a high degree of sensibility cannot exist in conjunction with a high degree of contractility is this:—the nerves (which are the seat of sensibility) are more or less actually sensible, according to the greater or less degree of firmness with which they are compressed on all sides by the parts immediately surrounding them and in contact with them. Thus the nerves of the bones, ligaments, and sinews are so firmly compressed on all sides by the unyielding structure of these parts, that they are almost wholly insensible. You may cut them, lacerate them, without giving pain. The muscles (that is, the red flesh) cannot be wounded without considerable pain, because their structure is not so firm as that of the bones, sinews, &c.: but it is much more compact and firm than the structure of the skin, and therefore a wound inflicted on a muscle will not produce anything like the acuteness of pain which is felt on wounding the skin. In the nerves of the eye and the ear it was necessary that a sensibility of the very highest degree should exist, in order to enable these organs to feel the very slight and subtle impressions of light and sound. Accordingly, we find that from these nerves all surrounding pressure is removed entirely; these nerves being, as it were, expanded into a sort of quivering jelly at that part where they are destined to receive their natural impressions.

Now, a very large portion of the body is, as you know, made up of a conglomeration of blood-vessels. The whole body, then, taken as a whole, will be the more compact and firm accordingly as these vessels are fully distended with blood;—precisely as sponge becomes a more compact body when distended with water than when dry; since, when dry, its cells are filled with air; but when saturated with water, they are filled with water, which is a far more compact material than air. If you draw a thread through a sponge saturated with water, the sides of that thread will be everywhere compressed and supported by either the solid parts of the sponge or by the water: whereas, if you draw a thread through a dry sponge, whenever that thread passes through an empty cell, its sides will be entirely unsupported and uncompressed. So of the body;—a nerve passing through the body (which body consists of a congeries of vessels) will have its sides everywhere compressed and supported, so long as those vessels are well filled and fully distended. But if these vessels be half empty—if their sides be allowed, as it were, to collapse and fall away from the nerves which they everywhere surround—it is manifest that those nerves will not be so firmly compressed and well supported, as they were while all the vessels surrounding them were fully distended. It is this half-filled state of the vessels which constitutes that lax and soft state of the body called flabby. This loose, flabby, and uncompact state of the body, therefore, is highly favorable to sensibility; since sensibility is always increased accordingly as surrounding pressure and support are removed from the nerves.

But a state of the body the very opposite to that just described is alone favorable to contractility: for I have before proved that vigorous contractility can only reside in recently organized matter: and in order that this recent organization—this perpetual youth as it were—of the several parts of the body may be kept up, an abundant supply of healthy blood is absolutely indispensable. Thus contractility in perfection requires a highly distended condition of the blood-vessels, and consequently, a firm and compact state of the body—a state exactly the contrary of that just described as favorable to the development of sensibility. Hence arises a most important corollary; viz., that whatever increases the natural vigor of the circulation, increases the contractility and lessens the sensibility of the body; and whatever lessens the sensibility of the body, by increasing its contractility, increases also the natural vigor of the circulation; since the blood is circulated by virtue of the contractile power of the heart and arteries.

The degree of sensibility, therefore, is always not only in an inverse ratio of the degree of contractility, but also of the circulating power: and since all the motions of the body are performed by virtue of contractility, and the whole process of nutrition by the virtue of the circulation, this is the same as saying that the degree of sensibility is always in an inverse ratio of the degree of health and strength;—which is the fact.

A very familiar instance of the increase of sensibility produced by lessening the quantity of the blood, is to be found in the fact, that a dose of cathartic medicine wholly incompetent to affect the bowels under ordinary circumstances, will be found quite sufficient to do so if administered after blood-letting.

I am at this time attending a huge, strong, dread-nought-looking custom-house officer, for a slight attack of paralysis which he sustained some weeks since. Since the attack I have bled him nine times, taking away thirty ounces of blood each time; he was also once cupped by Mr. Comley of Osborn street. You may easily imagine that a man who can bear this, and yet walk about the streets without support, must, at least, be no chicken. Yet so much has his sensibility—(I speak now of moral sensibility, which, after all, is but the physical sensibility of those parts of the nervous system which are susceptible of moral impressions)—so much, I say, has this man's sensibility been increased by bleeding, that a cross word is sufficient to make him burst into tears.

As moral sensibility is but the sensibility of those parts of the system which are capable of being impressed by moral causes, it follows that the qualities of the mind will be, in a great measure, regulated by the relative degrees of contractility and sensibility in individuals. When the brain and nervous system are but ill supplied with blood, and that blood but feebly circulated, and therefore imperfectly vivified, the sensibility to moral causes or stimuli will be morbidly acute. Such a person is easily and morbidly affected by causes to which others are wholly insensible: a sudden loud knock at the door, for instance, will make him start almost from his seat.

If you speak to him of a contingent evil, however slight and remote, he views it through a mental telescope, always applying that end of the instrument to his eye which magnifies the object and increases its proximity. If you speak of a contingent good, the telescope is instantly reversed; and he views it through the opposite end, which diminishes its value, lessens its probability, and renders it only visible at the extreme point of a long perspective. In short, he is timid, desponding, infirm of purpose, imaginative, and incapable of continued application.

Such a man may be a poet, but not a mathematician.

On the contrary, when contractility is vigorous, and the circulation consequently energetic, the brain will be abundantly supplied with healthy blood; its nervous tissue firmly supported everywhere within the meshes of that tissue formed by the interlacings of well-filled blood-vessels; and its sensibility, therefore, will be, in a corresponding degree, obtuse. It requires a strong moral cause to operate on the mind of such a man. "Trifles light as air" have no power to excite, to irritate, or in any way affect him: he is, consequently, bold, patient, good-humored, inflexible, unimaginative, and capable of long-continued mental exertion.*

Such a man may become a great mathematician, but never a poet.

I think I could show that all the peculiarities of the human mind are to be accounted for, as depending upon certain modifications of the two physical properties—contractility and sensibility; but on this subject I have said enough—and, perhaps, you will add, "and to spare:" therefore, my dear John, for the present, I bid you farewell. "*Pax vobiscum!*" E. JOHNSON.

* Why does exercising the memory strengthen the memory? Clearly, because it strengthens the brain by increasing the vigor of the circulation through it. The use of the hammer strengthens the blacksmith's arm in the same manner: viz. by increasing the vigor of the circulation through it; and consequently increasing also its contractility, that is, its strength.

LETTER V.

Umberslade Hall, near Birmingham.

MY DEAR JOHN,

I have now to speak of the principal actions concerned in the nutrition of our bodies—that is, in converting our food into ourselves. These are four in number: *absorption, circulation, respiration,* and *secretion.*

If you have read attentively what I have already written concerning the absorbent vessels, and concerning those arteries which perform the office of secreting the several juices of the body, as the saliva, &c.;—if, I say, you have read all this with attention, you will now have no difficulty in understanding what is meant by the terms absorption and secretion. I shall now, therefore, briefly describe the circulation of the blood, and the effect which respiration has upon it: and then I shall endeavor to exhibit to you these important phenomena of absorption, circulation, respiration, and secretion, in active operation, by tracing a given portion of food through all the changes wrought upon it, by virtue of these four actions, until it has become assimilated to the body.

First, let us trace the *circulation of the blood.*

The blood, of a bright vermilion hue, and richly laden with the elements of living matter—the new materials for repairing the wasted body—starting from the left side of the heart, enters the aorta. From the aorta it is distributed into branches of the aorta, and hence into branches of these branches, being divided and subdivided into smaller and smaller streamlets, as it proceeds from branch to branch. In this manner it is propelled onwards until it has been subdivided into as many minute hairlike streamlets as there are points in the body; there being no point of the body which is not supplied and nourished by one of these minute streamlets of blood.

While these countless myriads of currents are thus traversing the body—each, as it were, intent on reaching some one particular point or other as the end of its journey—they may be appropriately likened to an innumerable swarm of bees; each laden with stores, and hastening onwards in order to deposit its own particular share at some point or other of the honeycomb, which they are all mutually engaged in building or repairing.

When the blood has thus arrived at every point of the entire body, and each streamlet has fulfilled its office of renovation, by parting with the new materials which it contained, and depositing them in the place of the old and worn-out materials which have been removed, but the instant before, by absorption—when, in a word, the function of nutrition has been performed, the little hair-like arteries, which brought these several minute streamlets of blood from the heart to the several points of their destination, bend back upon themselves, lose the structure peculiar to arteries, assume that peculiar to veins, and commence their journey back to the heart.

The little streamlets of blood which fill these little backward-running veins having now parted with those living elements—those fresh materials—which they brought for the renovation of the body, may be likened, not inaptly, to the same swarm of bees mentioned before; which, having deposited their precious burdens in various parts of the honeycomb, are now hastening back for a fresh supply.

The blood, therefore, having fulfilled its functions, quits the arteries, and enters the veins.

I have said, that when the arteries cease to be arteries, and become veins, they bend back upon themselves. The veins, therefore, in their passage towards the heart, run alongside the arteries, and parallel with them; and wherever you find an artery bringing arterial blood from the heart, you will also find, by the side of it, and generally enclosed in the same sheath with it, a vein carrying back venous blood to the heart. Thus the several streams of venous and arterial blood pass each other on the road, as it were, like two trains of carriages moving side by side, but in contrary directions—the one train going out, the other returning home.

As the terminations of arteries form the beginnings of veins, it follows that the number of the veins, at their commencement, is equal to the number of the arteries. But these numerous minute veins, as they travel towards the heart, are every now and then uniting to form larger ones; consequently, the streams of venous blood, as they approach the heart, are constantly becoming larger and larger also; and thus the whole quantity of venous blood is eventually collected into two large veins, which empty themselves into the right cavity of the heart, like two Fleet ditches emptying themselves into the Thames.

We have now completed what is called "the great circle of circulation;" that is, we have traced the vermilion nutritious blood from the heart to every point of the body. We have seen it there part with its nutritious particles, in order to repair the waste of the body; and thus, deteriorated in quality, altered in color, and rendered oppressive and unwholesome in its properties, we have traced it back to the same organ from which it set out, viz. the heart. But, although we have brought it back to the same organ from which it started, we have not yet brought it back to the same side of that organ. It set out from the left side of the heart; and we have only traced it back to the right. Let us therefore proceed.

When the black deteriorated blood has been brought back from every part of our structure, and collected into the two large veins, which I have denominated "Fleet ditches," and has been poured by them into the right cavity of the heart, the walls of that cavity contract upon it, and propel it into a large vessel, termed the pulmonary artery, by which it is conveyed to the lungs. In the lungs, the pulmonary artery is divided and subdivided into an infinite number of minute branches, which traverse every part of the lungs. The black blood, therefore, carried to the lungs by the pulmonary artery, is divided into numberless minute streamlets, which are conducted, in every direction, through the lungs, by the innumerable hairlike branches of this artery.

The lungs are made up of a countless number of small cells, through and among which the little streamlets of black blood are of course conveyed: and every time we draw in our breath, these cells become filled with air; and the air which they then contain comes in contact with the little vessels containing black blood; and acting through the delicate coats of these, it operates those changes in the blood which it was sent to the lungs for the purpose of undergoing.

What the whole of these changes are, is not thoroughly understood: but this much is certain—that, whereas the blood enters the lungs of a black color, and in a condition unfit to effect the nutrition of the body, it no sooner becomes exposed to the influence of the air in the cells of the lungs, than it loses its black color, acquires the brilliant hue of vermilion, and becomes at once endowed with all the properties necessary to the nutrition of the body, and to the production or secretion of the several juices, such as the gastric, the pancreatic, &c.

The black blood, then, having been exposed, in the air-cells of the lungs, to the action of the air, and having been by it purified, reimpregnated with nutritious particles, and every way requalified to fulfil its appointed offices in the body, is collected into four veins, called the "pulmonary veins;" by which it is brought back to the left side of the heart, from which it first started. And thus the lesser circle of circulation has been accomplished, and the whole circulation of the blood completed.

Allow me to recur, for a moment, to the metaphor of the bees. I like it: it is a little fanciful, perhaps, but nevertheless appropriate, and not inelegant.

Consider the lungs, then, as a bed of sweet flowers, upon which a swarm of bees (the little black streamlets of blood) have settled. These bees, having laden their thighs with honey, quit the flowers, and, taking their flight through the garden-gate (the heart), pursue their way, by various routes (the arteries), in order to deposit their little burdens, and distribute them equally throughout the honeycomb; that is, the body. Having done this, they take wing once more; and returning in the same direction, but by different routes (the veins), they re-enter the garden (the heart), and again settle themselves down upon the flower-bed (the lungs), in order to collect a fresh supply of honey: that is, of nutritive property. Observe: a stream of arterial blood is a bee laden with honey;—a stream of venous blood, a bee despoiled of its honey.

Now if, as I hope, you have understood my former Letters, you will recollect that the old body is constantly being dissolved, and carried away, and emptied into the venous blood; and that the new materials afforded by new food are also emptied into the venous blood at the same place (viz. just before it enters the right side of the heart, on its way to the lungs), by the lacteal absorbents. The black blood, therefore, when it reaches the lungs, has, mixed up with it, a portion of the old body in a fluid state, called "lymph," and also a certain quantity of fresh nutriment, also in a fluid state, called "chyle." But the fresh nutriment (that is, the chyle) has not yet become blood: it is merely mechanically commingled with the blood. The effect, therefore, which the air exerts on the blood in the lungs, is not merely to revivify old blood, but likewise to convert the chyle into blood. This conversion of chyle into blood is called "*sanguinification.*"

There is another important office fulfilled by respiration; viz., the expulsion from the body of such portions of the lymph as are no longer fit to remain in it, in the shape of that watery vapor which we denominate "breath." The mouth, therefore, is a portal, through which you receive the materials for a new body, and also through which you blow away the worn-out materials of the old. Every time you breathe, you blow away a little bit of your nose, a little bit of your ears, a fragment of your eyes, a particle of your brain, an atom of your heart: in short, a part of your whole person. If you chance to be walking in the fields, a portion, mounting through the air, assists in forming the clouds; and again, descending in showers of rain, contributes its share towards the formation of the multitudinous ocean. Another portion falls upon the green herbage of the meadow, and constitutes a part of the nourishment upon which that herbage subsists. Thus, not only "all flesh is grass," but grass, also, is flesh.

I shall now endeavor to exhibit the principal actions concerned in the nutrition of the body, by tracing a portion of food through all the necessary changes, until it has ceased to be food, and has become an integrant part of yourself.

Let us suppose you to be in the act of despatching a hearty meal, consisting of animal food and various kinds of vegetables. You first introduce it into your mouth—with your teeth you masticate it—by means of your tongue, you roll it about your mouth. This rolling about brings it in contact with the several excretory ducts of the salivary glands; which open on the internal surface of the mouth, as we have before seen. These ducts, by virtue of their sensibility, become aware of the presence of a stimulus (the food). The stimulation which the food in the mouth exerts upon the ducts is propagated along them to the glands out of which they issue, and which glands are thus excited to an increased secretion of saliva. And this increased secretion of saliva is the first of that series of actions by which the nutrition of the body is effected;—and in this, the very first stage, you see exemplified those three important properties of which I have said so much in my last two Letters—*stimulation, sensibility* and *contractility:* for it is by virtue of their contractility that the arteries supplying the salivary glands with blood are capable of acting; that is, of contracting—and so of supplying the glands with blood, from which blood the saliva is to be secreted. It is by virtue of the stimulating property of the food that their contractility is roused into action; and it is by virtue of their sensibility that they are aware that a stimulus is acting upon them.

The nutritious bolus, then, having been thoroughly masticated and rolled about the mouth until it has been well mixed up with saliva, is, by a very complicated movement, mounted upon the back of the tongue, and by it jerked into the throat, by which it is propelled downward into the stomach. Its presence in the stomach stimulates that organ, as it stimulated the glands of the mouth; and a copious secretion of gastric juice (that is, stomach juice) is brought about in the same way as a copious secretion of saliva was effected by its presence in the mouth. But, as there are neither teeth nor tongue in the stomach, the food, when there, cannot be so readily and at once mixed up and kneaded, as it were, with the gastric juice, as it was by means of those instruments comminuted and commingled with the saliva in the mouth: it is not, therefore, subjected all at once to the action of the gastric juice, but gradually, layer after layer. While the nutritious bolus is circumvolving within the cavity of the stomach, the gastric juice, poured out from the sides of the stomach, above and around it, falls upon its surface. When its upper surface or layer has been sufficiently acted upon by the gastric juice—when, by virtue of the inherent properties of this juice, it has been converted into a semi-fluid peculiar to itself, and called "chyme"—it floats off and away from the rest, towards the lower part of the stomach, where it (the stomach) is united to the upper extremity of the bowels. The upper layer of the alimentary mass having been thus converted into chyme by the action of the gastric juice, and sent away from the remainder, the next layer becomes exposed to the action of this juice;—and having, like the first, become converted into chyme, floats away after it to the pylorus, that is, the lower extremity of the stomach. Thus, layer after layer, the whole mass eventually becomes changed from the nature of food into chyme;—gastric juice, during the whole time this change is going on, being poured out from the internal surface of the stomach upon fresh surfaces of the alimentary mass. The whole quantity is usually converted into chyme in about four hours.

Now mark! whatever has been the nature and kind of the food which you have eaten, however heterogeneous the several viands may be, they must all be reduced to this unique homogeneous semi-fluid, called chyme—they must all lose their own several natures, and take upon themselves the one sole nature of chyme, and so become chyme itself—before they can leave the stomach, and enter the bowels, in order there to undergo the next necessary change.

Now, if you have eaten any matters at your meal which are what is called "difficult of digestion," that is, which are not easily assimilated to the nature of chyme by the action of the gastric juice; when these matters become exposed to the action of the gastric juice, they will necessarily require to be exposed for a longer time than is natural, because of the difficulty which the gastric juice experiences in reducing them to chyme. It will be, therefore, a longer time before these float off from the surface of the alimentary mass, so as to leave the next layer exposed to the action of the gastric juice; and the under layers or portions of food, which are waiting for their turn to be exposed, will be kept so waiting longer than the wonted space of time. The consequence of this is, that they are kept waiting, untouched by the gastric fluid, until they begin to undergo those changes common to all vegetable and animal matter, when placed in a warm, moist, and confined situation; viz., fermentation;—the vegetable matter undergoing the acid fermentation; and the animal, the putrefactive. For it must be remembered, that the food in the stomach still continues to be food, still remains unaltered, still continues, therefore, to be obedient to the common laws of fermentation and putrefaction, until it has its nature and identity destroyed, and a new nature and identity bestowed upon it by virtue of the action which the gastric juice exerts upon it. It ferments and putrefies, therefore, in the stomach (if not acted upon by the gastric juice) as quickly as it would do, on a sultry summer's day, in a small pantry, with its windows and doors kept shut. And this shows you the reason why such matters as undergo putrefaction with the greatest rapidity, as some fish, and fresh pork, do not well agree with weak stomachs; for that which putrefies most rapidly in the pantry will do so in the stomach.

While, therefore, the indigestible matters are slowly submitting to the action of the gastric juice, the good and wholesome portion of the food is actually putrefying, and can, therefore, afford no nutriment. During the process of fermentation and putrefaction, moreover, as all the world knows, a number of fetid gases are given out: these poisonous gases distend the stomach, weaken its energies, oppress its sensibility, enfeeble its contractility, diminish the secretion of gastric juice, and, in a word, disturb, interrupt, and wholly overturn the whole process of assimilation in the stomach; and there is tumbled into the bowels—instead of a bland, smooth, homogeneous, healthy chyme—a filthy, fermenting, yeasty mass, smoking with offensive gases, and consisting of little else than sour vegetables and putrid meat: for the sensibility of the pyloric valve—of which I am to speak directly—is overcome by the oppressive influence and expansive nature of the gases which are distending the stomach. Is it possible, I ask, that healthy chyle and sound blood can be formed out of such a villainous compound of nastiness as this? But I have here described an extreme case.

As soon as this vile compound reaches the bowels, it will generally be expelled by them with violence: and this is the way in which bowel complaints are so often produced by some sorts of fish, and fresh pork, when eaten by persons whose stomachs are too weak to furnish a sufficient quantity of gastric juice to reduce them to chyme before they have had time to run into putrefaction; and the wind which such persons discharge by the mouth, after eating, consists of the offensive gases above mentioned. Strong, healthy stomachs, pour out their gastric juice so rapidly and abundantly, that the whole meal is reduced to chyme before the process of putrefaction has had time to begin.—Now let us proceed.

The food, having been properly acted upon by the gastric juice of the stomach, is now no longer food, but a bland, smooth, homogeneous semi-fluid, called chyme; which, quitting the upper part of the stomach, flows downward to the lower extremity—that part where the stomach is joined to the bowels. This junction of the lower extremity of the stomach with the upper extremity of the bowels is called "the pylorus:" and the pylorus is furnished with a peculiar valve, which accurately closes the communication between the stomach and bowels at all times, excepting when chyme is in the act of passing out of the stomach into the bowels. This valve is endowed with a singular and most beautiful kind of eclectic sensibility; which enables it to know, by the feel, whether the matters which come in contact with it be pure chyme or not; and nothing can enter the bowels from the stomach without coming in contact with it.

Now, let us suppose that a portion of food has been reduced to chyme, has flowed down to the lower extremity of the stomach, and has presented itself at the pyloric valve for admission through it into the bowels; and let us suppose that there is, floating on the chyme, a portion of food which has not yet been sufficiently acted upon by the gastric juice. I will tell you what happens. As soon as the pyloric valve feels the presence of the smooth and bland chyme, it instantly opens, and allows it to pass; but no sooner does the portion of food which has not yet been reduced to chyme attempt to follow, than the valve instantly closes the aperture, and refuses it admission. This particle of food must, therefore, return to the upper part of the stomach, to be again submitted to the agency of the gastric juice, before it can be permitted to escape from the stomach into the bowels. Is not this a beautiful exemplification of the importance of the sensibility of our organs?—and said I not truly, when I called it "our guardian angel?" For what is the sensibility of the pyloric valve, by which it is enabled to distinguish between perfect and imperfect chyme?—what is it, I say, but a watchman, a sentinel, posted at the entrance into the bowels, in order to watch over their safety; to see that nothing be allowed to enter which is likely to disturb or irritate them; to take care that nothing injurious, nothing offensive, nothing which may be in any way hostile to their safety, nothing, in fact, which has no business there, be permitted to trespass within the sacred precincts of organs so important to the health and welfare of the whole being, of which they form so vital a part?

That imperfectly chymified food cannot enter the bowels without injury to them is sufficiently proved by the very existence of this valve. For surely it is foolish to suppose that nature, who does nothing in vain, would have been at the pains of establishing so beautiful, so wonderful a contrivance, if the office which it fulfils were not in the last degree essential!

What mischief, therefore, do those persons inflict upon themselves—what a wide door for the admission of all sorts of evils do those persons throw open—who, perpetually stimulating the pyloric valve by the unnatural stimuli of ardent spirit and highly-seasoned sauces, enfeeble, wear out, and eventually destroy its sensibility: so that whatever the caprice of the palate throws into the stomach, is tumbled, right or wrong, assimilated or unassimilated, good, bad, and indifferent, altogether, without let or hindrance, into the bowels! for the sentry-box is deserted—the watchman is dead.

When I contemplate this state of things, I think I see a whole army of diseases marching in file out of the stomach, through the pyloric gateway, into the citadel of the bowels. I see pale-faced and bloated Dropsy with his swollen legs—livid Asthma struggling for breath—grotesque and tottering Palsy—yellow-visaged Jaundice—red-eyed Delirium—Fever, with his baked lips and parched tongue, looking piteously around, and crying, "Water! water!"—limping Gout, grinning with pain—musing Melancholy—hideous Insanity!—But let us drop the curtain over a picture so horrible. My mind's eye aches with looking at it. Above all things, my dear John, take care of your pyloric valve!

Now let us get on a step further.

The food having been thoroughly and properly acted upon by the gastric juice in the stomach, is reduced to a uniform substance, called "chyme;" and this is the first great change in that succession of changes which is ultimately to convert it into blood. There is now neither bread nor meat in your stomach—there is nothing there but chyme; which is neither meat nor bread; but a fluid, the nature of which is one degree nearer to the nature of blood than it was before it became chyme.

The chyme then flows to the lower extremity of the stomach; presents itself at the pyloric valve; and, having been examined, as it were, by the sensibility of that valve, and reported "all right," is admitted into the duodenum.

The first twelve inches of the bowels, reckoning from their junction with the stomach downward, are called the duodenum.

Now, the chyme in the duodenum has precisely the same effect upon the excretory ducts of the liver and pancreas, which open into the duodenum, as the food had, in the mouth, upon the excretory ducts of the salivary glands; that is to say, it stimulates the mouths of these excretory ducts; and this stimulation is propagated along the ducts to the glands themselves—the liver and pancreas. These glands, so stimulated, pour out an increased quantity of their individual secretions; viz., bile and pancreatic juice. The surface of the bowel itself, too, (the duodenum), pours out an increased quantity of fluid, called "the intestinal juice." The chyme mingling with these juices, another remarkable change is effected: the chyme is no longer chyme; it has lost its identity; and the result is, two fluids, of which one is called chyle.

This conversion of chyme into chyle, and another fluid,* forms the second great change, by which that which was once food (bread and meat), has been advanced two degrees more nearly to the nature of blood.

I hope you have not yet forgotten that the chylous absorbents arise, by open mouths, from the internal surface of the bowels. As the chyle, therefore, flows along the duodenum, it comes into contact with the open mouths of the chylous absorbents. These, by virtue of their sensibility, become aware of the presence of the chyle which is stimulating them to action. They answer the call, by erecting themselves, protruding themselves forward, dipping, as it were, their mouths into the chyle, and then retracting and closing them, thus performing an actual suction (if you will allow the term), by which the chyle is drawn within the calibre of these beautiful little vessels.

The chyle, thus absorbed, travels along the lacteals (that is, the chylous absorbents); is filtrated through their glands; is emptied into the thoracic duct; and by it, is poured into the blood of the veins about the bottom of the neck, and is carried by the current of the blood through the right side of the heart, along the pulmonary artery into the lungs.

While the chyle is traversing the chylous absorbents and their glands, it undergoes a change, the nature of which is not understood; but it is a change which advances it another degree nearer to the nature of blood. By the time, therefore, that your dinner —or, rather, that which was once the food which constituted your dinner—has reached your lungs, it has become almost blood; but it has not yet become quite blood.

When the chyle has reached the lungs, it is then exposed to the action of the air which we inhale, in the manner which I described when speaking of the circulation of the blood. Here the final change is effected; and that which was bread and meat has now entirely lost all its former characteristics. It was first food, then chyme, then chyle. Now it is none of these: it has acquired, by virtue of the agency of the air in the lungs, the color and all the other qualities and properties of blood: in a word, it has become blood itself. Thus, comparing the animal economy to the economy of vegetable life, one might say, that the stomach and bowels are the soil; bread and meat are the seed which is sown therein; and blood is the fruit which that seed produces—a fruit which is

* This *other* fluid is also absorbed into the blood by the veins.

destined to become the food of the animal. For, as was justly said by Hippocrates, "there is but one food, although there are several forms of food." However various the viands may be which we put into the stomach, they must all be converted into one and the same fluid, viz. blood, before they can have any effect whatever in nourishing or strengthening the body. Blood, then, is the sole nourishment on which we subsist; what we eat being no more than so much seed sown, with the view of producing a nutritious fruit, by which the body is to be fed, and its health and strength sustained;—viz. blood. We are no more nourished or fed by what we eat, than sheep are nourished by the turnip-seed which the farmer sows. The turnip-seed soon loses its identity; but in doing so, it gives rise to a turnip; and it is upon this turnip that the sheep feeds, and not upon the seed which produced it. And, in like manner, what we eat loses, like the seed, its nature and identity; but in doing so, it produces blood; and it is by this blood that our bodies are fed, nourished, and sustained. For, as the turnip is not the seed, but the product of the seed, so neither is the blood bread and meat, but the product of bread and meat; and it is from this new product that we derive our strength; and it is this which constitutes our food. Hence becomes manifest the utter impossibility of deriving any manner of nourishment or strength from substances which are incapable of being converted into blood; for example, ardent spirit;—no mechanism, no chemistry, no power, no magic, is capable of converting brandy into blood.

Hitherto, then, we have only seen the seed sown, and the proper fruit produced. We have now to mark the manner by which the body is fed and nourished by this fruit, viz. the blood. By the way, I may as well take this opportunity of calling upon you to take notice how little the quantity which we eat has to do with the quantity of nourishment which we derive from it; for, as the stomach, liver, &c. can only furnish at one time, enough of their several juices to convert a certain portion of what we eat into chyme and chyle, it is manifest, that only a certain portion can be converted into blood. And as blood is the sole aliment from which we can derive sustentation, it is equally manifest that we cannot derive any benefit from what we eat, except from that portion of it which in due course becomes blood. All that we eat, therefore, beyond what can be converted into blood, is either changed into that useless encumbrance called fat, or is left in the

4*

stomach and bowels to run into fermentation; serving no other purpose than to distend these organs with all sorts of pernicious and offensive gases.

In order to exhibit the manner in which the body is nourished —that is, the manner in which the fluid blood is converted into the solid parts of the body—it will, I think, be better to trace, to this consummation, only a single drop of blood at a time. You will, by this method, more readily understand it. But, by a drop I do not mean a great, round pumpkin of a thing, like a rain-drop, or a dew-drop; but a delicate minute globule, visible only to the eye of imagination—like the glow-worm's tear of disappointed love, when she lighteth her lamp in vain.

You have just seen the fresh chyle taken up by the chylous absorbents, and emptied, by the thoracic duct, into the veins at the bottom of the neck. Let us follow a single globule of this chyle.

Hurried along by the current of blood in these veins, it passes through the right side of the heart, along the pulmonary artery; then through one of its branches, into the substance of the lungs. Here it is acted upon by the air in the cells of the lungs, loses its characteristics of chyle, and becomes blood. It now turns round, as it were, and hurries back again out of the lungs along the pulmonary veins, to the left cavity of the heart.

But before we trace its progress any further, let us suppose that a hungry absorbent has just carried off a single particle from the point—the extreme protuberant tip of your organ of smell—"the very topmost, towering height o' Johnny's nose." The carrying off this particle would necessarily leave a little hole. Now let us go back for our globule of blood, which we have just traced from the lungs to the left cavity of the heart.

Rejoicing in its new existence, it leaps out of the heart into the aorta, hence into the carotid artery, thence into the external carotid, thence into the facial, thence into the superior coronary, and thence into a minute branch which the superior coronary gives off; which branch takes its course toward the tip of your nose.

By the time the artery, along which the little globule of blood is travelling, has nearly reached the tip of your nose (worthy to be called proboscis), it has become exceedingly minute, and its course tortuous; for it is now forming part of the ultimate tissue of the tip of the nasal promontory. The little globule, therefore, now moves along with diminished rapidity. Gradually it approaches nearer and nearer; and just when it has arrived exactly opposite to the little hollow left by the absorbent, becoming suddenly obedient to the secret agency of the nerves, its nutritious elements dart through the coats of the artery, like rays of light through glass, into that hollow, and at that instant become part and parcel of one of the most goodly noses within the four seas. The artery now turns back, soon loses the characteristics of an artery, and becomes a vein; by which vein the rest of the little globule is conveyed back, through the heart, to the lungs, there to be mingled with fresh chyle, and revivified by the action of the air in their cells.

This transformation of the fluid blood into the solid body is called solidification.

Now this is the way in which all the solid parts of your body are formed and maintained; every inch of it, therefore, once floated in your arteries, in the shape and quality of blood: and you see how foolish it is to suppose that there can be any real nutriment in those strong drinks to which the multitude attribute so many nourishing properties. What an inscrutably mysterious power, too, is manifested in this process! How wonderful, that so common and simple an affair as a potato should contain within itself all the elements necessary to the composition of an eye, an ear, or a tooth!—that this unheeded and unvalued root should be capable, within a few hours, of being changed, by commixture with the juices of the body, and by exposure to common air in the lungs, into blood!—and that from this single fluid, made out of this single potato, should be produced all those diversified and heterogeneous matters which make up the total of the body—the brittle bones, the soft and pulpy brain, the hard and horny nails, the silky hair, the flesh, the fat, the skin, the bitter bile, the sweet milk, the salt perspiration—everything in fact, from the corn on my lord's toe, to the down on my lady's cheek—from the sweat on the brow of Labor, to the dew on the lip of Beauty! Does it not seem incredible, that the ear, which can take cognizance of the faintest pulsations in the air, and appreciate with so much accuracy the value of musical tones—that the eye, wherewith the astronomer numbers the stars, taking in, at a glance, the half of heaven's whole orrery—nay, that the very brain, wherewith he thinks, and muses, and ponders over his problems and his logarithms and his equations—that the very brain itself of a Newton and a Shakspeare should own no better or nobler source than that of a despised potato! And, then, to think that that brain must die—must rot, and be resolved into its parent earth! Yet this is but the simple truth; and thus, like Ixion's, revolves forever the wheel of all existence—round, and round, and round—in an eternal circle of successive changes.

I shall now take leave to call your attention to certain facts which necessarily result from what I have said; and of which I wish you to take especial note.

First, then, you will observe, in following the food from the mouth, through all its intermediate changes until it has become blood, that almost all those intermediate changes are wrought upon it by the agency of the several fluids, juices, or secretions which it meets with in the mouth, stomach, and bowels; and that, consequently, its due conversion into healthy blood depends upon the healthy quality and abundant quantity of these secretions. But these secretions, like everything else in the body, are formed out of the blood; and their quality and quantity will, consequently, depend upon the quantity of vermilion blood wherewith the organs in which they are produced are supplied. And the quantity of blood with which these organs are furnished must depend upon the vigor and activity of the heart and arteries, whose office it is to convey it. Thus, then, it becomes clearly manifest, that a vigorous circulation is absolutely necessary to the assimilation (vulgarly called digestion) of our food. Whatever causes and habits of life, therefore, are calculated to give strength and activity to the circulation—as, for instance, exercise—is clearly of the first importance to the nutrition, and, therefore, to the health and strength of the body; and whatever causes and habits have a tendency to depress the energy of the circulation—to allow the blood to creep languidly through the body, instead of dancing along its channels cheerily and energetically—as, for instance, cushioned laziness, which rides when it should walk—must, of necessity, have the direct effect of impairing assimilation, and therefore of enfeebling the strength and sapping the very foundations of health.

But the energy of the circulation must exclusively depend upon the energy of the heart and arteries: and the energy of these, as has been already shown, depends necessarily upon the energy of their contractility; and energetic contractility depends on an energetic circulation, and is incompatible with a high degree of sensibility. Hence it directly follows, that whatever causes are calculated to increase sensibility—to make us tender, if you will tolerate a vulgar expression—have an immediate and powerful effect in impeding the conversion of our food into blood, and, therefore, of impairing the process of nutrition. Hence arise the incalculable mischiefs of a daily indulgence in what are miscalled the comforts of life; but which are, in reality, most pernicious and unnatural luxuries. A few of these are, table-indulgences, lounging on couches, warm carpeted rooms, window-curtains, bed-curtains, blazing fires, soft beds, flannel under-clothes (I speak of the healthy, not of the sickly invalid), novel reading, hot suppers, and, though last, by no means least, that precious piece of foolery, called passive exercise —that is, lolling along at ease in a stuffed and cushioned carriage. Not that I would totally abolish any one of these, except, perhaps, hot suppers and soft beds; but that I wish, by proving to you their evil influences, to induce you to use them as sparingly as the conventional habits of society will permit: though I confess, for my own part, I see no reason why any man should feel himself called upon to injure his health—to blur the beauty of God's noblest work—solely to gratify the capricious whim of that many-headed monster, that blatant beast, called *society*.

Again, the brain itself is the product of the blood—it is as literally and truly made of blood, as the most beautiful china vase is made of clay. Hence the qualities of the brain—the mental energies, as they are called—courage, the powers of abstract thinking, fortitude, patience, generosity, and, above all, good-humor,* can only exist in conjunction with, and owe their very being to, a vigorous circulation. Hence it seems scarcely too much to say, that thought itself is produced from the blood; since there can be no energy of thought without energy of brain, and no energy of brain without energy of circulation through that brain.

Thought is an act of the will. It is an act by which certain ideas are, to the exclusion of all others, summoned to present themselves to the mind's eye, that judgment may marshal them, compare them, and newly combine them. Thus, in solving a mathematical problem, the *will* suffers no ideas to intrude, save only the necessary ones of lines, angles, &c.

But the *will* is one of the energies of the brain; and we have

* If you go in search of good-humor, you must look to find it playing on the ruddy cheek and laughing in the unclouded eye of athletic strength. The sensibility of the athlete is too obtuse to be easily irritated. The skin of his mind is thick: and causes capable of excoriating others have only power to tickle the athlete.

just seen that these energies can only fully exist in conjunction with a vigorous circulation. When the circulation, therefore, is languid, the *will* will be languidly exerted—it will be unable either to command the presence of the ideas required, or to discard those whose presence is troublesome, and which only tend to perplex and interrupt the process of thought.

When a man with such a brain sits down to think, he finds that all sorts of ideas, wholly irrelative to the subject on which he wishes to think, are perpetually thrusting themselves into his mind "against the stomach of his will;" and so excluding those which a feeble and irresolute *will* is vainly endeavoring to summon and retain. If he be reading a book, he will find, every now and then, that though his eye has been tracing the words and lines, and his hand has been mechanically turning over the leaves—he will find, I say, that his mind has been wandering far away, and knows no more of what he has just been reading than the man in the moon. In a word, he has no power of abstract thought—no power to fix his attention. This state of mind is called reverie.

Herein consists the difference between thought and imagination. Thought, as I said before, is an act of the *will;* and that act, to be efficient, requires a vigorous circulation. It is the office of the *will* to decide, as it were, as to what ideas shall be admitted into the brain, and what refused admittance. But imagination resembles a dream, in which the *will* is asleep; it is a condition of the brain, in which all sorts of heterogeneous ideas, in despite of the *will*, come and go, in a tumultuous disorder, without let or hindrance, as in a dream. In this state of the brain, the contractility of its arterial tissue is feeble, and therefore the circulation through it is feeble; and therefore the *will*, which I have shown to depend on a strong circulation, is also feeble. In this state, the brain may be likened to an ideal theatre, without either check-takers or money-takers, and with all its doors thrown open, at which doors a multitudinous throng of ideas, of all colors and costumes, collected from all the corners of the earth and every domain of nature, are perpetually making their "exits and their entrances." And as the little pieces of colored glass in a kaleidoscope will often arrange themselves into figures more beautiful than any art can imitate, so, on the stage of this imaginary theatre, parties of these ideas will frequently frolic and gambol themselves into groups more grotesque, more picturesquely beautiful, than any effort of thought and judgment can accomplish.

Energy of will, therefore—firmness of purpose—the power of abstract thinking and reasoning—are all incompatible with a lively imagination; because the three former require an energetic circulation, while the last depends on a circulation of a contrary character.

There can be little doubt, I think, that insanity has its cause in some injury to the vigor of the circulation through some part of the brain.

That the doubts and fears and anxieties of the lover have a depressing effect on the circulation, is a fact long since established. The pensive dreamy sadness, the absent mind, the fondness for solitude, the long-drawn impassioned sigh so characteristic of love, is equally characteristic of a languid circulation.

The same condition exists in the poet; and the mental characters of all three will be found to possess no small similarity. So great indeed is this resemblance, that those who begin by being poets or lovers, not unfrequently end by becoming madmen. They are all three (generally) weak, wavering, wayward beings, incapable of abstracting their minds at pleasure, unable to control their thoughts;—and it may almost be said of all three alike, that they have scarcely any will or purpose of their own. Hence

> "The lunatic, the lover, and the poet,
> Are of imagination all compact;"

and hence it is true, that the poet does not sit down to think what he shall write, but to write what he thinks. But the word "think," in the last instance, is improperly used; he sits down in order to describe the ideas which his mind's eye beholds dancing in antic and ever-varying groups on the stage of his own brain's theatre—to

> ———"body forth
> The forms of things unknown;
> Turn them to shapes; and give to airy nothings
> A local habitation and a name."

Hence, too, every true lover is a poet, and every true poet a lover.

Finally, my dear John, you will observe that everything connected with life—all the actions, the energies, and beauties of the body—all the actions, energies, and beauties of the mind, as well as the body and mind themselves, are under the dominion of the circulation of the blood, from which both mind and body must inevitably derive each its tone and character. So that "the body and the mind are like a jerkin and a jerkin's lining;—rumple the one, and you rumple the other."

I have now described to you as much of the structure of the body, and its functions, as I conceive to be necessary, in order to enable you to understand what I have presently to say on the subject of diet and regimen. And you must now know quite enough to be heartily convinced of the unmitigated folly of those persons, who, without knowing anything of the structure of living parts, or of their actions, or of those delicate springs, contractility and sensibility, which originate and sustain those actions—who, I say, being as ignorant as idiotism of all that concerns the nature of life and living things, are nevertheless perpetually tinkering their stomachs with quack remedies;—thus stupidly presuming to mend a machine, of the very nature and structure and actions of which they are as uninformed as infant Hottentots.

The health of the body depends upon the healthy performance of the nutritive actions; and disease consists in the unhealthy performance of these actions, or of one or more of them. Medicines, therefore, with very few exceptions, such as those which seem to cure by chemically combining with and neutralizing the poison in the system which produced the disorder—medicines, with these few exceptions, have no power over disease, excepting as they have the power of increasing or diminishing the activity of the nutritive actions—absorption, secretion, circulation, &c.

When a man examines his patient, the question with him is not, Has he got a fever; or this, that, or the other disease? The question is, Which of the living actions is going wrong? and how is it going wrong? Is it going too fast or too slow? The patient has, perhaps, a foul tongue, dry skin, a quick pulse. But these are not the disease: these are the symptoms—the outward signs of the disorder within. He has nothing to do with these except as signs by which he ascertains the cause producing them. The question, therefore, is not what is good for a foul tongue, a hot skin, and a quick pulse: but what medicine possesses the power of controlling that particular living action—a disturbance in which has produced, in this particular instance, the symptoms in question. I say, in this particular instance; because, in others, the same symptoms will be produced by a disturbance in a different living action. The same symptoms, therefore, frequently require different treatment; because the cause of those symptoms is different, although the symptoms themselves are the same. I will give you a familiar instance. One man has a foul tongue, a quick pulse, and a dry skin, produced by inflammation of one of the membranes of his brain: he therefore requires leeches to his head. Another man has the same symptoms, from inflammation of the mucous membrane of the bowels: he requires leeches too—not to the head, but to the abdomen! Again, if a medical man finds his patient in pain, he does not forthwith run home for a dose of opium, because opium has sometimes the power of relieving pain; but he first ascertains which of the vital actions, which, being disturbed, is producing that pain. If it arise from spasm, opium may be of service; but if it arise from inflammation, opium will do harm, instead of good. If it were only necessary to attend to symptoms, and not to the cause of those symptoms, then the proper remedy for a foul tongue would be a scraper. One man has a headache from inflammation of the brain; another from flatulence of the stomach:—brandy will kill the one, and cure the other.

Again: cough may be produced by tubercles in the lungs—by inflammation of their mucous membrane—by inflammation of their coverings—by inflammation of their parenchymatous substance—by disease of the heart—by disease of the liver—by an accumulation of water in the chest—of matter in the chest, &c., &c.

I will tell you what happens every day. One of the faculty of ninnies gets a cough; and meeting with another, he is assured that so, or so, or so, is a "fine thing for a cough." The "fine thing for a cough" is straightway procured. Shortly, he has occasion to call on his tailor; and his tailor incontinently recommends him another "fine thing." The following week, his tinker brings home a mended saucepan; and then the tinker's "fine thing" must have a trial also. Then comes the butcher, and the baker, each armed at all points with "the finest thing in the world for a cough."

But, somehow or other, the cough still goes on—"ugh, ugh, ugh," barking away as before. Having frittered away a month or two in these follies, he then does just what he should have done at first—he walks off to the doctor, who finds that the cough was produced by inflammation of the covering of the lungs, which the abstraction of a little blood and a blister would, at the onset, have removed at once; but that, now, coagulable lymph has been poured from the inflamed surface, the covering of the lungs is adhering to the lining of the chest, and the patient has contracted a deadly disease, which no art can remedy. The tinker and the tailor, when informed of this, lift up their hands and eyes, and cry, "Dear me! who could have thought it?" and then march away to their other customers; to whom, if they happen to have coughs too, they very composedly recommend "their fine things for a cough" over again.

Is it not perfectly astonishing, that a carpenter, or a bricklayer, who would never think of pretending to mend your shoes, should, nevertheless, have no hesitation whatever in offering his services to mend your health? If you carry your kettle to be mended to any one but a tinker, he will tell you honestly that he knows not how to do it. But you shall travel from Dan to Beersheba, and, should you meet a thousand passengers by the way, not a soul of them but will undertake, should you complain of being unwell, to cure you on the spot.

Now, all this folly and mischief is attributable to no less a personage than that respectable old lady, said to be the mother of Wisdom—I mean Experience. It happens thus:—Mr. Noaks gets a pain in his bowels—his neighbor Styles experienced a similar pain last week, took brandy, and got well. Relying on this experience, he recommends brandy to Noaks. Noaks takes a glass, and feels better—another glass, and feels better still—a third cures him. Next year his son complains of a pain in his bowels; and his father, mindful of the experience of himself, and eke his neighbor Styles, administers to his son, in full confidence, a bumper of brandy. The son gets rather worse; but then his father recollects that the first glass did not cure his own pain, and so he gives his son another, and advises him to go to bed. Next morning, however, the pain being no better, some other neighbor assures the father that he has often experienced wonderful relief, whenever he has had a pain in the bowels, from gin and peppermint. So the father gives the son a bumper of gin and peppermint. But, although brandy, and gin and peppermint, might have cured the colic-pains of his two neighbors, it would not be found to be quite the thing for the inflammation which is already raging among his poor son's bowels. At last the doctor is called in, who finds that his patient has been laboring for thirty or forty hours under a disease which will often kill its victim in twenty-four; and that however mild it might have been at its onset, it has now, by the aid of brandy and gin, been urged on to incurable violence.

Experience may be the mother of Wisdom, for aught I know; but she is certainly the mother of Mischief also. Experience may teach a man to make bricks, and to lay bricks; but, without the scientific knowledge necessary to inform him how to use her, she can never teach him the practice of physic. Money is of no use to a man, unless he knows how to lay it out: and experience is unprofitable, unless a man knows how to apply it. And as money may be laid out to the injury of the spender, so experience, misapplied, becomes a curse in the hands of its possessor. Farewell!

E. Johnson.

LETTER VI.

Umberslade Hall, near Birmingham.

MY DEAR JOHN,

I have now given you what I hope you have found to be a tolerably clear notion of the several actions concerned in nutrition —and of the nature of life and health.

I have now to point out to you what I believe to be the chief causes and sources of disease.

When a man, who thinks as well as sees, suffers his eye to range over the various minor systems which compose the one great scheme of the universe—when he looks at the planetary system, and beholds worlds whirling amid worlds in countless numbers, with inconceivable rapidity, yet infallible precision—when he dwells on the vegetable system, and sees myriads of plants rising from the same earth, living in the same air, warmed by the same sun, watered by the same rain, yet each differing from each, and affording year after year, forever, each its own peculiar product, with unerring exactitude—the vine the grape, the oak the acorn, the brier the rose, the foxglove its bells of blue, the holly its berries of red;—when, with more inquisitive glance, he penetrates the thicker veil with which nature has curtained the chemical world, and watches the several phaenomena resulting from chemical operations—combustion, putrefaction, vegetable fermentation, &c., and observes the unfailing exactitude with which all these render obedient homage to the one great law of affinity;—then, when he looks inward, and contemplates his own system—beautiful as the most beautiful, and not less worthy of Omnipotent Wisdom than the most worthy—when he looks inward, I say, and beholds there all confusion and imperfection—when he perceives that, of all the systems of nature, that of man alone is liable to derangement, and is the only one of all which ever fails of fulfilling its intentions—when he sees that while all others always go right, his own goes almost always wrong; when, moreover, he reflects that his own system is the work of the same Almighty hand which fashioned and gave being to all the others—when the eye remarks all this, the mind cannot but be irresistibly struck with the anomaly; and the tongue cannot but exclaim, "Why is this so? How is it that the system of man—of man, the master-miracle of creation—how comes it that the system of man is forever going wrong, while all around him goes right? The natural average of human life, we are told on high authority, is "three-score years and ten." How happens it, then, that "about one-fourth of the children that are born, die within the first eleven months of life; one-third within twenty-three months; and one-half before they reach their eighth year?—that two-thirds of mankind die before the thirty-ninth year, and three-fourths before the fifty-first?—so that, as Buffon observes, of nine children that are born, only one arrives at the age of seventy-three; of thirty, only one lives to the age of eighty; while, out of two hundred and ninety-one, only one lives to the age of ninety; and, in the last place, out of eleven thousand nine hundred and ninety-six, only one drags on a languid existence to the age of a hundred years. The mean term of life is, according to the same author, eight years in a new-born child. As the child grows older, his existence becomes more secure; and after the first year, he may reasonably be expected to live to the age of thirty-three. Life becomes gradually firmer up to the age of seven; when the child, after going through the danger of dentition, will probably live forty-two years and three months. After this period, the sum of probabilities, which had gradually increased, undergoes a progressive decrease; so that a child of fourteen cannot expect to live beyond thirty-seven years and five months; a man of thirty, twenty-eight years more; and, in the last place, a man of eighty-four but one year more. Such is the result of observation, and of calculations on the different degrees of probability of human life, by Halley, Graunt, Kersboom, Wargentin, Simson, Deparcieux, Dupre de St. Maur, Buffon, D'Alembert, Barthez, Dupre, and M. Mourgues."—*De Lys' Richerand.*—How is it, that of the whole number of children, so few, so very few, live long enough to fulfil the final cause of human existence?

Now if, in contemplating the system of man in connection with the other systems of nature, we be able to discover any one very striking difference wherein his system differs from all others, may we not fairly presume that this difference between them is the cause of the remarkable and otherwise unaccountable anomaly above mentioned?

We need not look far, nor ponder long, in order to discover the difference which distinguishes the system of man from that of all others;—and it is indeed a momentous one! It is this: that while all the other systems of the universe are sustained and governed by immutable laws, as gravitation, chemical affinity, instinct, &c., &c., the system of man depends solely for support upon laws, the perfect or imperfect fulfilment of which has been left dependent on the capricious conduct of man himself. For the laws which sustain the human system are the laws of nutrition; and these are forever subject to disturbance by man's misconduct. For instance: a man may voluntarily half starve himself—or by his folly he may bring himself into a position in which he is unable to procure sufficient food—or he may take greatly too much—or he may select for food such substances as are incapable of being assimilated to his own structure—or he may annul the laws of nutrition entirely by taking aliment of a poisonous quality. Now, it is perfectly manifest, that under any of these circumstances the laws of nutrition must be seriously modified—injuriously disturbed. And it is equally clear, that these circumstances, in the instances supposed, are the result of human conduct. The systems of the lower animals are also sustained by the same laws of nutrition; and these laws are also liable to be modified by the conduct and habits of these animals: but, then, the conduct and habits of brutes are themselves dependent on instinct, which is unerring: whereas the conduct and habits of man depend on his own caprice—the use or abuse of his reason, which is not unerring.

The grand distinction, therefore, between all the other systems of nature and that of man seems to be, that while the former are sustained by unerring laws, the latter is supported by laws which are subservient to the erring conduct of man, with relation to the manner of his nutrication* and mode of existence.

Now, it is to this "erring conduct with relation to the manner of his nutrication and mode of existence," that I look, as the cause and source of human disease.

To every system nature has allotted a determinate position: and she has established a fixed relation between each system, and all the other systems by which each is surrounded; and from their allotted position none can swerve—their own allotted relation to

* Be careful not to confound the term "nutrication," with the term "nutrition." Nutrication signifies the supplying the mouth with food; nutrition, the assimilation of that food to our own structure.

surrounding objects none can disturb—none, except man. But man, as I hope to prove to you hereafter, has removed himself from his natural position—has broken down his natural relation to the external world; and so brought himself within the sphere of the operation of causes injurious to his well-being, which could not otherwise have reached him.

All planetary phenomena, as we have just seen, as well as those of brute life, of chemistry, of vegetable life, mechanics, and physics in general, owe their infallibility to the infallibility of the laws which sustain them: and I think it cannot be doubted, that the fallibility which distinguishes the system of man from all others, has its origin in the fallibility of the laws on which it depends for support, or rather the fallibility of that conduct and mode of existence on which those laws depend for their perfect or imperfect fulfilment. If the immutable law of gravitation—which, as it were, bridles the planets, guiding and restraining each in its proper path—had to depend for its energy and constancy upon the caprice of man, is it not easily conceivable—nay, is it not absolutely certain—that the system of the planets would have been liable to as many disorders as is the humano-animal system? Should we not speedily have had a repetition of those scenes, in which the North Pole glowed with summer heat—when the lazy Bootes ran sweating away with his wagon, and the Moon could not but express her astonishment on seeing her brother's curricle and four in the very act of trespassing on her own highway?—should we not have had hot fits and cold fits—fevers and agues—disordered functions and diminished secretions? Would not the Moon occasionally have forgotten her function of reflection, and the Sun his secretion of light?

By a parity of converse reasoning, had the system of man been made to rely for its sustentation on some immutable law—like that of gravitation—had the nutrication of the body been effected by some invariable law over which man possessed no control—had he himself nothing to do with the feeding his body, and had he possessed no power to alter his allotted position and relation in the universe—in a word, were we fed by chemical affinity, and held in our places by some physical law—then the actions which constitute the life and the health of the human machine would have been as unerringly executed as the revolutions which constitute the health and the life of the planetary scheme. A well-constructed watch, if properly defended from external injury, will indicate the hours of the day as infallibly as the Moon will revolve in her orbit in her given month: so, also, under like circumstances, would those movements and revolutions of the fluids which constitute the life of the human machine be executed with the same unfailing precision, provided only that the law of nutrication be properly fulfilled, and its proper position among the other systems of the universe duly observed. All things were created with a view to the fulfilment of a final cause; and it is insulting to the Creator to suppose that He has attempted to attain to a final cause by means which are inefficient for its accomplishment.—But it may be denied that other systems are infallible. It may be said, that there are occasionally certain signs observable in the heavens which seem to indicate that then and there a world has gone to pieces. Be it so. But who shall say that it has gone to pieces before it has fulfilled its final cause—before it had existed its appointed term! I am not attempting to prove that man is not "born to die"—I am only endeavoring to show that he was not by nature subjected to disease and premature death. I claim for the system of man no more than is readily conceded to other systems. I claim for him only the same degree of perfection, the same importance, the same consistency, which are so clearly observable in all the other works of the Almighty Architect of the universe. I cannot believe that it formed a part of the original scheme, that one half of mankind should die before they have attained the age of eight years—that is, before they have lived long enough to fulfil any one conceivable intention—in fact, before they are themselves fully formed.

If any man die while any one of his organs is unimpaired, he dies prematurely, and before he has fulfilled the final cause of his existence. For nature is an economist in everything: she creates nothing in vain: she never falls short, nor does she ever exceed the object in view: she husbands her resources, and never wastes her energies. But to create an eye or an ear with the power of seeing or hearing for eighty years, and to attach that eye or that ear to a body capable of existing only sixty years, would be an obvious waste, a most unnecessary expenditure of energy. This would be like loading a blunderbuss to shoot a sparrow. What would you say to that architect who should employ fifty men for fifty days in erecting a column of stone to support a bird-cage or a pepper-box? The means, my dear John, which nature employs, are always exactly proportioned to the end—not an atom too little, not an atom too much.

If this reasoning be not admitted, then we are driven to the conclusion, that the human system contains within itself, as part of its primitive design, and wholly independent of man's conduct, the principles of disease and premature death. But that *some* individuals do escape both these—both disease and premature death—the evidence of our senses daily assures us. In these individuals, therefore, either these principles do not exist, or they exist to no purpose. These principles, therefore, can only form a part of the primitive design of some individual systems; or, if they do form a part of the original scheme of all, they are clearly only effective in some. But, surely, to suppose this, is to make such a haphazard affair of human life—is to convert this "harp of a thousand strings" into such an ill-contrived and discordant kettle-drum—is to reduce it to a thing of such mere contingency, that no one but the infidel proselyte to the doctrines of blind chance could reconcile it either to his reason or his conscience to believe it.

But that disease and premature death formed no part of the original design of man, is superabundantly proved by the innumerable contrivances which nature has instituted, in every part of the machine, to repel them; and the mighty efforts which she makes, under disease, to escape them.

My inference, then, is this: that the vital actions constitute a system of nature, which is, like her other systems, perfect in itself;—that, as the planetary system depends for its health (that is, the due performance of its functions) on the law of gravitation, so the health of the vital actions depends on man's conduct, with relation to the manner of his nutrication and mode of existence; and that, as the planetary system is incapable of derangement while the law of gravitation remains unchanged, so neither is the system of man capable of disorder, otherwise than by some misconduct in his manner of feeding and his habits of life. Beyond the influence which results from our conduct in these two respects, we possess no more control over the motions constituting health, than we do over those of the heavenly bodies; everything being effected by the inherent powers of the nutritive system itself, by virtue of the laws which govern that system: and to suppose that, while all other systems are fulfilled of necessity, the system of human nutrition is fulfilled *par hasard*, and may or may not answer its intention, just as it may happen, is to suppose that which is in direct opposition to the evidence of our senses, as it regards the uniformity, simplicity, and perfection of nature; and is, therefore, directly opposed to right reason and common sense.

The instance of hereditary diseases does not invalidate this argument: because, although the inherited disease be not contracted by any error of diet and conduct in the inheritor, yet, I contend, it must have been originally derived from such a source, to the parent who first became the subject of it. For instance: a man, from high and gross feeding, contracts gout; his sons, however temperate, may nevertheless be afflicted with gout by inheritance; that is, supposing gout to be an hereditary disease, as some assert. Here you see, notwithstanding the temperance of the son, his gout was evidently the result of error in diet: not, indeed, on his own part, but on the part of his parent. And it must be remembered, that I am speaking, not of individual disease, but of disease in general.

A child may be born with some imperfection in one of the valves of the heart: but this imperfection is the result of some imperfection in the action of those vessels whose office it was to form this valve: and this second imperfection could only be derived from some imperfection in the health of the parent, induced by the causes in question. Death from dentition, again, is the result of a morbid irritability, produced partly by the imperfect health of the parent, and partly by the operation on the infant of the same causes which enfeebled the health of its parent; viz. improper diet, and improper habits.

I know there are a few diseases which result from climate, situation, soil, &c.; but these are so few, as rather to prove the rule, than overturn the argument.

What I wished, therefore, to prove, and what I hope I have proved, is, that disease and premature death formed no part of the original design of man: and that for the long funereal list of disorders to which we are subject, with the exception of a very few, we are indebted solely to ourselves.

It seems to me, that there is but one legitimate cause of death; and that is, old age;—and here, as ever, nature shows herself a kind and watchful mother. There is nothing painful in death from old age; it makes its advance with a gradual and stealthy step, which is scarcely noted; and the old man drops into the tomb almost insensibly; conscious, indeed, that it cannot be far distant, but still ignorant of the moment when it may open to receive him. By imperceptible degrees, the living principle becomes more and more feeble; the heart's pulsations less and less frequent; the fluids circulate with diminished rapidity—a change is wrought in their quality; they perform their several offices imperfectly; the food is slowly assimilated; we have bone where we ought to find cartilage; we have flaccidity where we ought to find firmness and tension; bones, which before were separated, now become consolidated; the fluids lubricating the joints are deficient; the ligaments regulating their extent of motion are indurated. Thus, the old man moves with difficulty, and his respiration is hurried and unequal on very slight exertion. The least essential parts of the body forsake him first: his hair becomes white, and falls off; the teeth loose, and drop out; his vision becomes impaired, his hearing imperfect, his judgment inaccurate, his temper querulous: a little while, and he becomes perfectly helpless; his brain loses its sensibility; his memory deserts him; already the twilight of death is around him; and shortly the night of the grave closes over him, and he is no more seen. Lastly, comes Oblivion with her sponge, and wipes his name from off the tablet of human recollection; and the bustling hero of this little drama is heard of, and thought of, and finally even dreamed of, no more.

I lay this down, therefore, as a fundamental truth, that we bring disease upon ourselves, by using an improper diet, and by exercising improper habits of life; and that the only way to preserve vigorous health, and strength of mind and body, is to bring ourselves to the use of a proper diet, and the exercise of proper habits of life, as nearly as the tyranny of custom will admit. But, before we can do this, we must first ascertain what diet and habits are those which are proper to man.

Now, my dear John, when the mother of a newly-born infant dies, the physician recommends a wet-nurse for the nourishment of the infant. In the absence of a wet-nurse, he recommends that ass's milk should be given to it:—in the absence of this, cow's milk;—in the absence of cow's milk, he recommends mild farinaceous food. He does *not* recommend cold boiled beef and ale.

But why does he not recommend cold boiled beef and ale? Because he knows that these are unnatural food for newly-born infants.—Why does he recommend mild farinaceous food, in preference to beef and ale? Because this, though not the natural food of infants, is, nevertheless, not so unnatural as beef and ale. But why does he prefer, for the infant's support, cow's milk before farinaceous food? Because cow's milk is more naturally allied to the food of infants than farinaceous food; that is, it approaches, in its own nature, more nearly to the mother's milk than farinaceous food does.—But why does he prefer ass's milk to cow's milk? Because ass's milk is still more naturally allied to an infant's natural food than even cow's milk.—But why does he prefer a wet-nurse to all the other means of nourishment? Because human milk, though not the mother's, is the most natural of all. But why does he recommend that mode of nourishing the infant which is the most natural? Because he believes that that mode which is the most natural is also the most proper.

Thus you see that the physician, in selecting the food most proper for a newly-born infant, takes no other guide than Nature herself. Uninfluenced, in this case, by his own appetites, passions, and prejudices, he sees with a clear vision, and understands with an unclouded mind. He does not hesitate for an instant—the question requires no time for consideration—it is so simple and so manifest that it will not bear a doubt, or an argument, for a moment. "Human milk," says he to all inquirers, "is the most natural food of infants, and *therefore,* beyond all doubt, question, or cavilling, the most *proper.*" He would laugh in the face of any one who should dare to question this fact. "Nature," he would exclaim, "is the highest of all authorities—it is the authority of God himself." Yet only turn the tables, and inquire of this same physician, what he conceives to be the diet most proper for himself. Only mark! how instantly the case is altered. See how he shifts, and turns, and doubts, and quibbles! How he endeavors to distort or discolor truth, and twists sense into nonsense and nonsense into sense. His vision is no longer clear, his understanding no longer unclouded. His perception of truth is instantly obscured by the mists which are exhaled from his own passions, personal prejudices, and appetites. He so loves his soup—so delights in a ragout! A bumper of hock is so refreshing, and port is so invigorating! They make him so happy—so comfortable—he cannot believe they are injurious—at least, not in his own particular case! He no longer takes nature for his guide—her authority no longer suffices—and her voice, which, in the case of selecting the infant's food, was the voice of God, becomes, in his own particular case, the voice of a lying spirit. Pitiful sophister! miserable prevaricator! as though Truth could change her essence, and Nature her immutable laws, in order to tickle the palate and accommodate the propensities of a wretched gourmand.

Surely, if that food which is unnatural to infants is also improper to infants, so also must that food which is unnatural to man be also improper! And if this be true of food, must it not also be true of habits? If nature be a paramount and indisputable authority, must she not be so in one instance as well as another?

Look through the universe—not at this or that particular part of it, but look everywhere; search minutely through all the kingdoms of nature; explore the natural world; examine curiously the artificial world—whatever you behold, whether animate or inanimate, moving or at rest, large or small, you will find it has been placed in a sphere of its own, and surrounded by circumstances peculiar to itself; from which sphere and circumstances it cannot be removed, without detriment to the integrity of its natural perfection. In fact, all things—the very stocks and stones—have "habits" proper to themselves; and you cannot compel them into new habits, without injury to their primitive perfect condition. Everything has its determinate position, and fixed relation to all other things. It is this which constitutes that wonderful harmony which so astonishes and delights those who love to contemplate the works of nature. If, indeed, it were not so, nothing but the most inextricable confusion must necessarily be the immediate result.

Everything, then, has its natural sphere of existence—its natural habits; and you cannot compel it out of its sphere, without injury to the perfection of its being. I know I may be asked, if the marble, chiselled into the statue, be not an improvement upon the rugged mass? I answer, No—decidedly not. The marble is not improved: it has been made to contribute to the enjoyment of man, it is true; but this is improving the condition of man, not the condition of the marble. For, in the first place, the marble itself is unchanged, except in figure; and it has been wrested from the security of concealment in its quarry, and exposed to injuries and accidents from which it would otherwise have been exempt; its very existence as marble has been rendered precarious;—a barrel of vinegar may be spilled upon it, and so its very nature be changed, and its identity destroyed. In the next place, looking upon the whole quarry as one object, of which the statue forms only a small part, and supposing (as who shall dare to question it?) that the entire quarry was intended by nature to answer some useful purpose in the general scheme, I ask, has not the capacity of the entire quarry, to fulfil its allotted purpose, been diminished by the forcible abstraction of a part of it? If you fell a mahogany tree, in order that it may be wrought into billiard-tables, and side-boards, and dining-tables, I ask you again, have you committed no injury upon that tree? Have you abstracted nothing from the beauty of that scene in which that tree made a prominent object? Have you in no way interfered with the purposes for which that tree had its being? Or will you choose rather to suppose that nature planted mahogany trees for the express purpose of veneering side-boards and dining-tables? Did they answer no intention, did they effect no useful object, before these same side-boards and billiard-tables were invented? "Oh! but," says man,

"All things were made for my use."

We have such a consummate opinion of our magnificent *selves* that whatever we find capable of being made to contribute to our own enjoyments, we instantly conclude, with a pompous vanity, not a whit less ridiculous than that of the frog in the fable, was made and sent expressly for our own behoof.

With what a proud sense of superiority do we look down upon the inferior animals!—yet how slight an accident is sufficient to degrade the most towering genius beneath the level of the most inferior? A stone in his path trips up his heels; or a little tumor forms upon his brain; or a few ounces of water collect between its membranes or in its ventricles; and, behold, the vaunted philosopher, the lord of the creation, has suddenly become a drivelling idiot! "Toi, qui dans ta folie prends arrogamment le titre du roi de la nature—toi qui mesures et la terre et les cieux—toi pour qui ta vanite s'imagine que le tout a eté fait, parceque tu es intelligent, il ne faut qu'un leger accident, qu'un atome deplacé, pour te de grader, pour te ravir cette intelligence dont tu parois si fier!"*

But let us admit for an instant, that all this were so—that nature planted mahogany trees on purpose to veneer Crockford's Rouge-et-Noir tables; this detracts not an iota from the truth of what I have asserted; because it must still be admitted, on every

* Translation.—"Thou! who in thy folly arrogatest to thyself the title of the King of nature—thou! who measurest both the earth and the heavens—thou, for whom thy vanity makes thee imagine that all things were make because thou art intelligent—there needs but a slight accident—but a single atom displaced, in order to degrade thee, and ravish from thee that intelligence of which you seem so proud!"

hand, that the tree, as a tree, "has had foul wrong." No one can deny that a tree, which has been cut down and cut up piecemeal, has suffered injury as a tree—has had the integrity of its perfection, as a tree, destroyed. My assertion, therefore, is still sound—that you cannot withdraw any object from its natural sphere, without detriment to that object.

Here is a limestone:—it would have remained perfect limestone, probably forever, had it been left in its natural position, the quarry. But I have withdrawn it from its natural sphere—I have broken its natural relation to surrounding objects—I have thrown it in the fire, and exposed it to a shower of rain—and, behold! it crumbles into dust.

Here is a "winking Mary-bud:"—had I left it in the field whence I abstracted it, it would have gone on winking as prettily as any Mary-bud of them all; but I have planted it in tallow, and bathed it in ink, and, behold, it is dead! Poor flower! How piteously thou lookest—dropping ink, instead of dew, upon thy greasy bed! I would not serve another so to enlighten the darkness of fifty brother Johns!

Here is a watch:—I wear it in my fob—I place it beneath my pillow, or in my bed-room watch-pocket—and it never fails to indicate the time. But if I attach it to a mill-sail, or conceal it in an oven, or bury it in an iceberg, what sort of time will it keep?

Go visit the Zoological Gardens—observe the extreme care which is found necessary, in order to keep the animals in health. And in what does this care consist? Manifestly in approximating their present circumstances, as nearly as possible, to the circumstances in which nature intended them to live. Yet, with all their care and extreme attention, there are still animals which they have not yet been able to preserve alive—much less in health! You cannot withdraw the leopard from his jungle and his tropical climate, and turn him adrift on the plains of Siberia, with impunity to himself; nor will the cedars of Lebanon flourish on the barren hills of the frozen north.

As it is with habits, so with food. All animals cannot subsist upon the same food indifferently; nor all vegetables upon the same soil. A dog will not thrive on oats, nor a horse on beef, nor a cat on green gooseberries.

Seeing then, that every other system in the universe has its natural sphere of existence—its natural habitudes, from which it cannot be removed without injury—it seems only in accordance with strict analogy, to suppose that man also has a sphere of existence and certain habitudes natural to himself; and that he cannot withdraw himself from this sphere and these habitudes, without injury to the perfection of his nature.

What is the conclusion to be drawn from these premises? It is this: that "that manner of nutrication, and that mode of existence," which is most natural to man, is also most proper to man.

E. Johnson.

LETTER VII.

Umberslade Hall, near Birmingham.

MY DEAR JOHN,

In my sixth Letter I showed you that those habits and that diet are most proper for man, which are most natural to him. I illustrated this doctrine (with regard to food) by instancing the case of a physician when called upon to give his advice as to the best manner of nourishing a motherless infant. I showed that, in such cases, he takes nature for his exclusive guide, and, without hesitation, recommends a wet-nurse. And why? solely because human milk is the natural nutriment of infants, and *therefore* the most proper. And if this be so with regard to diet, so is it also as regards habits and general regimen. The wisdom and provisions of nature are the wisdom and provisions of God; and shall the capricious inventions of art be put in competition with *His* contrivances? It may be said that the ingenuity which teaches man the practice of arts was given him also by God. But who does not see the weakness and fallacy of such reasoning? Who does not see that, if this mode of reasoning were true, it would apply, with equal force, in favor of the ingenuity which teaches a man to pick his neighbor's pocket, or swindle him of his property by the cunning art of forging his signature?

Well, then, my dear John—in order to ascertain what are the habits and diet most proper for man, we have now to inquire what are the most natural to him.

This, dear John, is a curious subject, and it embraces a most extensive field of philosophical speculation: it is, nevertheless, one concerning which I would have you entertain right notions; as otherwise, a good portion of my time and labor have been thrown away. Since, unless you can be made to comprehend what is most natural to you, you cannot possibly understand what is most proper for you.

You cannot take for your guide the opinion of medical men. For if you consult twenty, you will hear twenty different opinions;

and which of the twenty can satisfy you that his, and his only, is the correct one? Neither do I ask you to rely on mine. I would have you trust to the *ipse dixit* of *none*. I would have you governed by the evidence, and guided by the light, of your own reason alone. For this is not a matter of speculative science, nor of normal art. It is a matter, simply and exclusively, of common sense.

It has been asserted that man has no natural condition; but that, into whatever state he brings himself, and with whatever circumstances he surrounds himself by the exercise of his ingenuity, that state, and those circumstances, constitute the sphere in which he was designed to exist. The same opinion has lately been expressed by a Medical Author, in one of our periodicals, with respect to diet.—"In short," says he, "whatever kind of food the ingenuity of man has been able to discover, that kind of food is proper for him." If the phrase had been, "the right exercise of his reason," I should have agreed with him; but the term "ingenuity" embraces, not only the use of reason, but also its abuse.

If this doctrine be sound, there is no article of diet, however notoriously unwholesome, which may not claim to be considered as the proper food of man. And if it be asked, "Why?" the answer is ready:—"Because it was discovered by man's ingenuity."

The ingenuity of the poor old women who earn a miserable subsistence by selling apples in the street, has discovered that the cravings of hunger may be more deeply allayed by a glass of gin than by bread and meat;—*therefore*, it was the design of Nature, with regard to these poor souls, that their appetites should be so appeased! Monstrous!

To me it appears, that man was designed to exist in that condition which the right exercise of sound reason shows to be most consistent with his nature.

The design of nature, with regard to the proper mode of existence amongst brutes, is rendered evident by her having bound them within the limits of their proper sphere by the force of an irresistible instinct. But to man was given reason, in the place of instinct; leaving him to use it or abuse it, at his option; otherwise she would have defeated her own design of making him a free agent.

It seems to me, therefore, that man has a natural condition; viz. that which a right exercise of sound reason leads him to adopt, as being most agreeable to his nature. He has also, therefore, natural habits, and a natural manner of nutrication; viz. those which right reason points out, and well-founded experience approves.

I ask, is the present condition of society—are the present habits and manners of man—natural to him? This is to say, are they the result of a right exercise of sound reason? Are they congenial with his nature? Or do they result from the abuse of his reason—from pampered passions, meretricious appetites, and vicious propensities?

No one will deny, that the ultimate aim of all men's pursuits—the final goal, towards which all mankind are running, or fancy they are running—the philosopher's stone, of which all are in search—the *ultima linea* of all human hopes and human efforts—is, *happiness.*

The right exercise of a sound reason, therefore, would certainly induce mankind to choose that path which experience has proved will conduct him to happiness, and would warn him against those which the experience of ages has shown will not lead him to happiness.

The question is thus brought to turn upon this single hinge: Has man, as he now exists in a high state of refinement, chosen the right path? Does it conduct him to happiness? Or, admitting that perfect happiness is incompatible with a sublunary existence, does it ensure him the largest possible proportion of happiness of which his terrestrial existence is susceptible? If not, then he has not chosen the right path—he has not exercised a sound reason in his choice; inasmuch as he has chosen to travel in a road which will not conduct him to the object of his journey.

To prove that man, in a high state of intellectual culture, is not happy, really seems a work of supererogation;—it is only to iterate the most universally-acknowledged of all truisms; it is merely to prove that which no one thinks of denying. Throughout all highly-civilized societies, and in all cultivated languages, the unhappiness of man has ever been a standing proverb in the mouths of all men, and a fruitful theme of declamation and lamentation, both with the philosopher, the worldling, the poet, and the preacher.—"All is vanity and vexation of spirit!"

A state of discontent is unquestionably a state of unhappiness:—and, contrariwise, I think you must be compelled to admit that contentment is happiness: for it is clear that he only can be contented who is in possession of the entire sum of all that he desires. But the entire sum of all that man desires is happiness. It is a logical *sequitur*, therefore, that he only can be contented who is in possession of happiness; and he only happy who is contented:—and this makes happiness and contentment synonymous terms.

There can be little happiness, therefore, where there is little contentment.

But look through society at large, as at present constituted. Do you observe content or discontent to be its grand characteristic? It is idle to deny, for it is impossible to conceal, that universal discontent is one of its most prominent features. From the monarch to the menial, "*nemo contentus*"—none are contented; and therefore none are happy. Our whole lives are consumed in the pursuit of an unattainable object.—What is that object? Happiness.—But why is it unattainable by us? Simply, because we are running after it, along paths which do not lead to it.

Look again through society; and observe our innumerable Institutions for the relief of human misery—our Hospitals and Dispensaries—our Philanthropic Institutions—our Asylums for the Destitute—our Penitentiaries—our Mad-houses—our receptacles for reformed prostitutes, foundling children, and other unfortunates! Can that be a happy or natural state of things, which makes necessary such institutions as these? It is ludicrous to hear people boast of these establishments, as so many proofs of the blessings conferred by civilization; whereas they are, in fact, standing monuments of its shame and disgrace. Is it a credit to us, that we live in a state of society in which sickness and suffering are so rife, that thousands of public Hospitals and Dispensaries are required to administer relief to the diseased? Is it a credit to us, that destitution is of such frequent occurrence, that public buildings are necessary to receive the destitute? Is it a credit to us that we live in a state of things, the tendency of which is so frequently to overturn the human reason, that it has become necessary to provide public buildings for the reception of the mad? Can that be a meritorious condition, which makes it necessary to provide a Mendicity Society in order to restrain beggary; and a Humane Society for the prevention of suicide?

Yet, once again, look through society. Look at our hosts of lawyers, and others engaged in the law—all of whom live, and thrive, chiefly from the proceeds of vice, dishonesty, and crime!

Look at our police-station-houses, our spunging-houses, our tread-mills, our prisons, our hulks, our convict-ships, and our colonies for the reception of transported felons! Look at all this;—and then deny, if you can, that "there is something rotten in the state of Denmark!"

There is one more contemplation which must, I think, carry weight with it, in showing how clearly and immediately crime and misery result from an artificial state of society. One of the first and greatest essentials to a highly-civilized condition of society is money, and if you reflect, for a few minutes, how innumerable have been the crimes, and consequent punishments and sufferings, which are traceable, in one way or other, up to money alone as their cause, the contemplation will be found to be perfectly appalling!

It is clear, therefore, that the present condition of highly-civilized man is not conducive to happiness; but that, on the contrary, it is the prolific parent of multitudinous misery. It is not, therefore, a condition which the right exercise of sound reason has led him to adopt, or which experience has approved: *therefore,* it is not his natural condition—it is not in accordance with the design of Nature—it is not the condition in which he was intended to exist.

We have just seen that the present condition of man is not consistent with his moral nature, inasmuch as it is not conducive to his happiness. Let us now inquire whether it be in consonance with his physical nature: that is, whether a high degree of civilization and refinement be conducive to his physical health and strength.

If you survey the several grades of society, you will find that the class of men who enjoy the highest degree of health and strength is precisely that which is the farthest removed from a high degree of refinement; it is that which approaches the most nearly, in its habits and condition, to primeval simplicity: I mean, the tillers of the soil—the agricultural laborers. This alone is surely a strong argument that the habits of refined society are not friendly to health and strength. Again: as a proof that a high degree of civilization is hostile to health, are not our numerous and crowded Hospitals, our multitudes of thronged Dispensaries, and our countless multitudes of medicine venders, and medical practitioners, quite sufficient?

Man, in a state of primeval simplicity, inhabiting the temperate latitudes, living almost entirely in the open air, supporting life by the simplest fare, his mind undisturbed by the harassing anxieties consequent on ambitious pursuits, and the thousand other perturbing causes inseparably connected with a highly cultivated state of society, enjoys almost an entire immunity from disease. But, from the moment he begins to emerge from the primitive simplicity of his habits, and seeks to live by his wits rather than by the sweat of his brow, from that moment his intellectual and physical energies are at perpetual war with each other; since he can only increase the former at the expense of the latter. As he advances in refinement and knowledge, he retrogrades in physical strength. And, to me, I confess, this fact alone would be an unanswerable proof, that a highly intellectualized state of society, like that in which we live, was never designed for man. It seems to me insulting to the wisdom of the Creator, to suppose that it should be so. If it had been intended that man's chief care should be the culture of his mind, it seems to me, I repeat, most insulting to Omniscient Wisdom and Omnipotent Power, to suppose that He would have so constituted him, that the very means which he must use to cultivate his mind are such as he cannot adopt without injury to his physical health, and even considerable risk to life itself.

It is certainly a natural law, that man shall preserve his health; —this being neither more nor less than a part of the law of self-preservation. But if it be also a natural law, that man shall cultivate his intellect, then this absurd consequence arises; viz. two natural laws, obedience to one of which necessarily involves an infraction of the other:—for nothing can be more notoriously true, than that the close confinement, sedentary habits, and perpetual tension of the mental faculties necessary to study, and the cultivation of the intellect, are highly detrimental to bodily health. We know, too, that feebleness of body has a direct tendency to enfeeble the mind: thus, the same causes which directly enfeeble the body have the effect, indirectly, of enfeebling the mind. The very means, therefore, which are necessary to educate and polish the mind, have also a strong tendency to injure it. "Intellectual cultivation," says Dr. James Johnson, "sows the seeds of physical deterioration;—and the evils thus inflicted on the flesh fail not to grow up, and ultimately retaliate, with interest, on the spirit."

I am not singular in my opinion, that a high degree of civilization is inimical to health. Dr. Southwood Smith, in his Philosophy of Health, says: "The usual, the permanent, the natural condition of each organ, and of the entire system, is pleasurable." And, commenting on this passage, Dr. James Johnson observes: "This might be true, if we were in a state of nature; but in our present condition, there is scarcely such a thing as perfect health." Elsewhere the same able author, Dr. J. Johnson, observes: "The great evil—the root of innumerable evils—the Proteiform malady, *Dyspepsy*—the hydra-headed monster of countless brood and Medusa mien, is the progeny of Civilization." In another part of his "Œconomy of Health," he says, and most truly, that the "besetting sin of the present generation . . . . is that of reading and thinking."

The immense consumption of drugs is another strong proof of the sickly health of society in general, in its present boasted era of refinement. I am sure I am within the boundary of truth when I assert, that, throughout England, there is not more than one man in a hundred who does not find it necessary, at least once in the year, "to take medicine," that is, to carry the master-piece of God's creative wisdom to the doctor, to have it mended.—Why, I would discard my tinker if my saucepan required mending so often!—"There are many millions in this country," says the author just quoted, "to whom physic is, daily, as indispensable as food!—To the luxurious epicure it may seem incredible, that within the boundaries of the British Isles there are thousands, among the opulent classes, who would give half their wealth to be able to do without food altogether—who would gladly give up the pleasure of eating, for an immunity from the misery of digesting." —Again, he says: "The state of civilization at which we have arrived produces such a wide range of 'hopes deferred,' and expectations blighted, that their effects are detected, by the experienced eye, at every step, even in the streets."—Again: "The results are read, by the observant physician, in the countenance, the complexion, the gait, the whole physical and moral constitution of the female;—results which require a new vocabulary; and would be totally unintelligible by Celsus, or even by Sydenham, could they rise from their graves, to survey the progress and effects of Civilization."

If further proof be wanted, to show that a highly-cultivated condition of society is prejudicial to health, we have it in the very structure of his body, and in the economy of its living actions. I have already shown you, that a vigorous circulation is necessary to vigorous contractility—that is, health and strength; and that vigorous contractility is incompatible with a high degree of sensibility.

No one, I think, will venture to deny, that cultivation and refinement have the direct effect of heightening man's sensibility. The very word "refinement" seems necessarily to imply a superior degree of sensibility. One can hardly conceive a high degree of refinement unaccompanied by a high degree of sensibility also; for the very meaning of the term "refinement" seems to be a condition from which everything calculated to offend a delicate sensibility is removed.

Besides, one of the first effects of civilization is, to substitute the labor of the brain for the labor of the hands and feet. But the labor of the hands and feet—exercise—is necessary to the existence of a vigorous circulation and an energetic contractility; which are both incompatible with a high degree of sensibility. Exercise, therefore, has the effect of blunting the sensibilities; and, by a parity of reasoning, a deficiency of exercise has the contrary effect—that of heightening the sensibility of man. "This deficiency of exercise in the open air," says Dr. James Johnson, "may be considered as the parent of one half of female disorders, by multiplying and augmenting the susceptibilities to all external impressions! The pallid complexions, the languid movements, the torpid secretions, the flaccid muscles, and disordered functions (including glandular swellings), and consumption itself, attest the truth of this assertion."

The substitution, therefore, of mental for bodily labor, which is one of the very first effects of civilization, manifestly tends to heighten our sensibility. "The superior cultivation of intellect—now so eagerly aimed at, as the means of rising in the world, indeed of getting through it—renders the feelings more acute, the sympathies more active—the whole moral man, in short, more morbidly sensitive to moral impressions. These impressions are annually multiplying in number and augmenting in intensity. The principal sources from whence they flow, in a thousand streams, on suffering humanity, are these:—the fury of politics; the hazards and anxieties of commerce; the jealousies, the envies, the rivalries of professions; the struggles and contentions of trade; the privations, discontents, and despair of poverty;—to which might, perhaps, be added the terrors of superstition, and the hatreds of sectarianism."*

* Œconomy of Health, p. 94.

In another part of the same excellent work, the author observes: "Thus, then, a nervous temperament, a morbid sensibility, pervades the whole frame of society, more or less—a super-sensitiveness, that inflicts pains and penalties on trifling and occasional indiscretions."

It is true that refinement multiplies our pleasures, but oh! by how much the more does it multiply our miseries! It is true that it multiplies wants, the gratification of which is productive of much delight. But who would be at pains to catch the itch, solely that he might enjoy the luxury of scratching?

Now, if it be true, that civilization and intellectual cultivation have the effect of raising the sensibility of man beyond the natural standard, then it cannot be denied that they are also prejudicial to his health and strength; since I have proved that energetic contractility, which is but another term for health and strength, cannot exist in conjunction with a high degree of sensibility.

Thus, then, it appears, that the present superlatively intellectualized state of refinement is an unnatural condition; inasmuch, as it is one, which a right exercise of sound reason, and the authority of experience, unite to prove to be neither in conformity with the moral, or physical, or structural nature of man. And I have already shown you, that whatever is unnatural is also improper. His present condition, therefore, is not his proper one.

But if the present condition of man be unnatural and improper, what state is that to which right reason and experience would point as the natural and proper one; that is to say, as the one most in accordance both with his physical and moral nature?—From what has gone before, the answer is manifest; viz., that state in which high civilization and excessive intellectual cultivation do not exist: —in a word, a state of patriarchal simplicity. It would be easy to show that the patriarchal condition is also most suitable to man's moral nature; that is, most conducive to his happiness. But this does not properly belong to my subject; which has for its object his physical health and strength only.

Now let us shortly recapitulate.

In my Sixth Letter, I endeavored to show that the human system, like all the other systems of Nature, has a natural determinate position, and natural habits; and that if it be suffered to remain in its natural position, and in the exercise of its natural habits, it will be found to be, like them, perfect in itself. I also endeavored to show, that if any of the other systems of the universe be removed from their natural position and natural habits, it cannot be done without injury to them.

In this Letter, it has been my object to show that man has been removed from his natural position, and from the exercise of his natural habits; and that from this cause arise the disease and misery to which we find him subjected. And the arguments which I have used to prove this have been, at the same time, sufficient (at least in my estimation) to show what his natural position and natural habits really are; and therefore, also to show what habits are most proper for him—most in accordance with his nature—and, therefore, best calculated to secure to him the perfect enjoyment of health, strength, and happiness.

The knowledge of what those habits are, which are most proper for man, must constitute your rule of conduct with regard to your own. For instance, if you be convinced, from all I have said, that the primeval condition is the one best calculated to secure health and strength, you need no other guide than that conviction. You have only to reflect on what that condition was—its habits and diet; and then endeavor to reduce your own, as nearly as conventional customs will allow, to that standard. Let your habits be more hardy—your fare more frugal. Eat only that you may live; and do not live only that you may eat. Diminish the activity of the mind, and increase that of the body. Work more and think less. Avoid the excitement of music, cards, wine, assemblies, politics, religious controversies, &c.; or share in them with the utmost moderation. The only legitimate excitement is that of exercise in the open air.

Let me not hear you tauntingly inquire, in common with those fools who mistake derision for argument, and a sneer for wit, and who are fain to cover their own folly and ignorance with the cloak of unmeaning sarcasm—let me not hear you inquire, with such small fry as these, whether I would have society go back to barbarism, live in the woods, and feed on acorns? I have nowhere expressed or hinted at any such foolish desire; nor have I given the slightest cause for such a question. By all that I have said as to the evils arising from a high degree of civilization, this has been my sole object, viz., to place in your hands an infallible test by which you may prove—an accurate gauge by which you may measure—the propriety or impropriety of any one habit or article of diet, concerning the healthfulness of which you may wish to be informed. The traveller in the desert, the prairie, or the wilderness, who takes the Polar star for his guide, does not expect to scale the heavens and reach the star itself. He finds it, nevertheless, an unerring guide towards the object he has in view. Your object is health. Take the manifest ordinances of Nature for your Polar star, and you shall surely reach it.

I know, my dear John, very well, that the general tenor of this Letter is so much at variance with the preconceived opinions of mankind, that those who are too lazy to think for themselves, and those who think in chains, and those who are afraid to think, and those who know not how to think, will not stop to ask themselves whether it can be true; but taking it for granted that it must be false, because in opposition to the general opinion, will pass it over as a piece of mere extravagance. I write not for such as these. But to you—who are not, I hope, of their number—I say, that a proposition being opposed to the general opinion forms no argument whatever against it; because there is hardly any man so ignorant as not to know that there is scarcely any one well-ascertained truth which was not once in opposition to the general opinion. Leave the general opinion, then, to those who, rather than examine closely into things, are content to take them for granted:—but you, dare to think for yourself;—and, in doing so, rest not satisfied with a shining surface, but look through and beyond the surface. I want *you* to look through the gloss, and the glare, and the glitter, and the gingerbread gilding, wherewith Civilization, like a painted courtezan, carefully conceals her deformities. I want you not to swallow the gilded nut whole; but to crack it, in order that you may see the rottenness and bitterness which lurk within. Depend upon it, the Refinement of which we make so loud a boast, is no better than a cheat—a smiling impostor, who comes to us with a wreath of roses round her brow, and Pleasure's wine-cup in her hand, while she conceals poison and the dagger beneath her spangled robe.

I am, dear John, yours truly,

E. Johnson.

LETTER VIII.

Umberslade Hall, near Birmingham.

My dear John,

In this letter I shall endeavor to give you a practical instance of the manner in which the present habits of society act in the production of disease—their *modus operandi.*

There is a condition of the body in which no actual disease—at least, no denominated disease—can be said to exist; and yet in which the whole of the living actions are deteriorated—in which no one individual organ can be pointed out as the seat of disorder, but in which all the organs perform their offices in an irregular and unhealthy manner, and that without any evident assignable cause. No part of the machine goes right. In the morning, instead of waking refreshed with new vigor, and ready at once to enter with alacrity on the business of the day, the patient is oppressed with a painful drowsiness which he finds it almost impossible to shake off. When he has, at length, dragged himself out of bed, he moves about with a feeling of weariness, greater than he felt when he went to rest. Tired, languid, and lazy, he feels as though he could almost give half he is worth for one hour more of sleep. His tongue and mouth are either parched, like the surface of dried toast; or foul, clammy, and exceedingly disagreeable to himself. He is unable to eat any breakfast; but is glad of a cup of tea or coffee, to cleanse his mouth and throat: and this is no sooner swallowed, than he begins to be annoyed with acid eructations, and perhaps sickness. In an hour or two he feels better, and gradually improves till dinner-time. At dinner his appetite is capricious: sometimes he can eat heartily; at others, scarcely at all. First, he finds one article of food disagree with him; then another; until, at last, there is scarcely one article of diet which he dares take. After dinner, lassitude and drowsiness again attack him; and he falls asleep, or sits gaping in his chair, till teatime, unless imperative necessity impels him to action. After tea, he again feels better:—and, indeed, from tea to eleven or twelve o'clock is the only time in which he can be said to be himself. His nights are passed either in a deep and dreamless sleep, or, more properly, stupor; or else he is restless and watchful; and has his short snatches of sleep perturbed by frightful dreams, and hideous nightmare. After slight exertion he feels disproportionate fatigue; and after a slight meal, he feels as though he had eaten too much. There is a constant sense of want and sinking, or faintness, about the region of the stomach, which frequently induces him to take a glass of wine or spirit; and the sudden, but delusive, and temporary relief which he experiences from this is, I believe, one of the most frequent incentives to habitual dram-drinking—another of the fatal effects often resulting from too implicit a reliance on experience.

He is almost ashamed to consult a medical man, for he scarcely knows of what to complain: he accuses himself of laziness—he drenches himself with physic: he is sometimes inclined to believe that it is all fancy, and he determines to fight against it, and to eat and drink like other people, and think no more about it. It is generally in the evening, when he feels a great deal better, that he takes this doughty resolution:—but it will not do. The morning comes, and with it the feverish tongue, the lethargic drowsiness, the weary limb, the languid spirit, the lassitude and listlessness, which makes his life literally burdensome. He is, indeed, in a miserable condition. Food does not strengthen him—sleep does not refresh him—mirth does not cheer him—society has no charms for him—pleasure no allurement: for him everything seems to have lost its interest. He can think of nothing but himself: his own wretched feelings are perpetually soliciting his attention, and forcibly abstracting his mind from every other contemplation. Like the owl, he mopes all day, and is only aroused into active existence at night;—and even then, should he dare to suffer himself to be tempted to indulge in a glass of wine, or a slight supper with his friends, he is haunted the whole time, and his comfort poisoned, by the dread of additional sufferings to be endured in the morning. He becomes hipped, nervous, melancholy, desponding. If his friends are not perpetually sympathizing with him, he fancies they have no regard for him. If they be merry, he imagines that they are exulting over him. He feels every smile as a personal cruelty, and the voice of mirth rings in his ear like the voice of the death-bell. His friends appear to have forsaken him*—Heaven itself seems to

* "Is it not strange," exclaimed the elegant but dyspeptic Rousseau, "that

have forgotten him. Thus apparently abandoned and neglected, can we wonder that the weight of all this wretchedness is more than he can bear; and that he finally welcomes suicide as the only possible relief from an intolerable load of imaginary oppression.

There are two other states of the mind produced by this unhappy condition of body, which are very singular. The one is a sense of dread and terror, as though the patient had committed some great crime: the other is an unaccountable and almost irresistible desire to do something horrible and wicked. I have seen numerous instances of these states of the mind, manifestly depending solely upon the condition of the body I have just attempted to describe. Of the latter—an irresistible longing to do something wicked—two cases occurred to me lately. A tailor awoke his wife at midnight, in great terror, and earnestly besought her to get up and put out the night-light; otherwise he said, he felt that he must set fire to the house. He had been lying awake, he said, looking at it for the last two hours, and had restrained himself with the utmost difficulty from using it to fire the tenement. The feeling had gradually been growing stronger and stronger, until he felt it had become irresistible.

In the other case, a girl, about seventeen years of age, was drinking her tea before the fire. There was a large looking-glass over the mantel-shelf. Suddenly she exclaimed, in a frightened tone of voice, "Oh, mother! mother! for God's sake, take the cup and saucer out of my hand!" She did so, and asked what was the matter. The girl drew a deep breath, and said, she did not know; but all at once she felt, that if the cup and saucer were not taken from her, she must have thrown them at the looking-glass over the mantel-shelf. She dared not, for weeks afterward, look at that glass, while she had anything in her hand. Both patients were in a very ill state of health.

I have no doubt that many of the extraordinary cases of shop-lifting, of which we hear, result from the same physical causes—a desire to do something wicked, without object or motive of any kind, depending solely on a morbid condition of the health, a prurient curiosity to see whether they *could* do it undetected, a morbid and itching wonder, and entirely unconnected with moral depravity. The *opportunity* to steal breeds the *desire* to steal; and while speculating within themselves as to whether they could

all the world should be leagued together to oppress the son of a poor watch-maker."

do it without discovery—while saying to themselves: "Now I could do it—now—and now"—a sudden and hurried impulse prompts the act, and thus the thought becomes father to the deed.

These are some of the characteristics of that bane of artificial life, commonly called "indigestion." Another group of symptoms indicative of congestion—congestion of the *capillary arteries* of the great nervous centres—a *blood-shotten* condition of the nerves and nervous centres—consists of extreme excitability, a disposition to give the greatest importance to the merest trifles, indecision of character, irritability of temper, inability to read or write, for any considerable time together, or to concentrate the mind and fix the attention, sense of weight and oppression on the top of the head, constriction across the brow, constipation, flushing of the face, especially after eating, terror on being left alone, indisposition to go into society, or to mix with strangers, distension of the stomach and bowels, sickness and headache, foul tastes in the mouth, flatulence, failure, in some shape or other, of the eyesight, sense of great heat on the top of the head, agitation and palpitation of the heart, waking out of sleep with a sense of terror, inability to bear rapid motion on horseback, or great exertion of any kind—these form another group of symptoms which too often harass both the mind and body of the poor dyspeptic. The whole of these symptoms may not always be present at one time, nor do they always exist with the same degree of intensity; but they are, more or fewer of them, and in a greater or less degree, its unfailing characteristics: and there are few persons, excepting those who earn their livelihood by bodily labor, who do not occasionally fall into this condition to a greater or less extent;—and not a few have their whole lives embittered by it.

Now this, my dear John, is that condition of body, to avoid or to remedy which is the object of all dietetic rules; for it is that from which almost all other diseases spring. If from having neglected this previous state of things you become affected with any distinct disease, as fever, cough, pain, &c., the best and only advice I have to give you is this: "Go at once and put yourself under the care of that medical man—(not who wears the smartest coat, drives the smartest equipage, and trims his whiskers after the smartest fashion; nor who smiles with the sweetest grace, speaks in the kindest tone, has the whitest hand, and the softest voice)—but whose general conversation bespeaks him a man of talent—a scholar—and, not a coxcomb, but a gentleman. In choosing a medical man, and in endeavoring to satisfy yourself as to his talent, you should not question him on matters connected with his own profession; because you are no judge therein—you have no means of knowing whether he answers you wisely or foolishly; but talk to him on matters of general learning and philosophy—on subjects with which you are yourself acquainted: you will thus easily fathom the depth of his understanding: and you may be quite sure, if he show that he possesses a mind capable of reasoning on the principles of science in general, that he cannot be ignorant of that particular science which he has made his especial study. For the practice of medicine is not an art, but a science: it cannot be learned by rule or by rote, as bricklayers learn to lay bricks, or a tallow-chandler to make candles. In the treatment of almost every disease, there is much to which no rule will apply; and the correct or incorrect management of which must entirely depend upon the philosophical powers and general reasoning capabilities of the mind of the practitioner who has the treatment of the case under his care.

Now let us inquire, as to what is the actual state of parts—the actual condition of the solids and fluids of the body—in these distressing circumstances of the health. I believe it to consist in sanguineous congestion in the ultimate tissue of all the organs concerned in the nutrition of the body—and, as a necessary consequence, in a deficiency and depravity of all the necessary secretions.

Everything tends to prove that man was destined to lead a life of bodily action. His formation—his physical structure generally, and that of his joints particularly—his great capacity of speed and laborious exertion—the divine injunction that he "shall live by the sweat" (not, mark you! of his brain,) but "of his brow"—the circumstances under which he first appears upon the earth—the bodily imbecility and enfeebled health invariably consequent upon a sedentary life—all go to prove that he was destined to lead a life of physical activity—of hardihood and exertion—and not of ease and luxury, nor excessive *mental labor,* which is *incompatible* with much and continued physical exertion. The circulation of the blood, by which all the actions concerned in nutrition are effected, is carried on with an increased rapidity under bodily exertion. If, therefore, a state of physical activity be natural and proper to man, so must that rapid condition of the blood's circulation be natural and proper also; because the one is a natural consequence of the other, and the former cannot possibly exist without the latter being produced. The *natural* pulse of man, therefore, is the *rapid pulse of bodily exertion.* From this it follows, that the degree of velocity and vigor with which the blood flows through the body, during inaction, is preternaturally diminished: one of the essential means destined to propel it has been withdrawn; and a too languid circulation is a necessary result. But, as during that increased rapidity of circulation consequent on exertion, there is also an increased secretion and excretion of perspiration and pulmonary halitus; so, when the circulation is languid, these are deficient: and as these are separated from the blood, so, when they fail to be separated, the greater must be the volume of blood remaining. Thus one of the principal natural means for reducing the blood's volume being removed, there must be accumulation somewhere; and as the larger arteries are not permanently dilatable while the veins and capillary arteries are so, this accumulation or congestion must take place in the veins and capillary or hairlike *arteries.*

The blood is propelled through the ultimate tissue chiefly by the contractile power of the heart acting upon the blood, as it were, from behind. When this power, therefore, is but feebly exerted, it is manifest that the blood will not be driven through the ultimate tissue with the requisite degree of velocity. Under these circumstances, the blood creeps sluggishly, and, as it were, lazily along the minute vessels composing the elementary tissue: they become gorged; and this engorgement operates as a still further impediment to the free flow of the blood.

But there is another most important evil resulting from this semi-stagnation of blood in the ultimate tissue. Arterial blood, when not moving with the due degree of velocity, becomes deteriorated in its properties:—for if you enclose a living artery between two ligatures, the blood so insulated assumes the black color and other properties of venous blood.

The blood, therefore, when not circulated with sufficient energy through the ultimate tissue, becomes deteriorated in quality—and this, too, precisely where it is of the utmost importance that it should be of the very highest degree of purity:—for, as you now know, it is in the ultimate tissue of our organs that all those operations are effected on the blood on which the nutrition of the body depends. The blood, when the circulation is driven on with a due degree of healthy vigor, maintains its vermilion hue and arterial character, not only as far as the extremities of the hairlike arteries, but even for some way along the beginnings of the veins. But when the propelling power from behind—that is, the power of the heart—is not sufficiently energetic, the circulation through the elementary tissue is so slow, that the blood loses its vermilion hue and arterial characteristics before it has reached the extremities of the hairlike arteries; and thus that part of the tissue which ought to be filled with arterial blood is gorged only with venous blood, from which the proper secretion necessary to the nutrition of the body cannot be separated, either in due abundance, or of a healthy quality. All that blood which, under exertion, is driven to the surface of the body, to the skin, to the muscles, and along the superficial veins, is, during inaction, sleeping and creeping, in muddy and sluggish currents, along the tortuous and delicate vessels composing the ultimate tissue of the internal organs—impeding their action, distending their coats, and disordering their sensibility.

The whole circulating system may be divided into three portions: the heart and large arteries, whose office is merely to convey the blood to the elementary tissue;—the elementary tissue itself;—and the large veins, whose office is merely to convey the blood back to the heart. Now, the principal force by which the blood is propelled from the heart through the whole of these three portions of the system, back to the heart again, is the contractile power of the heart itself. When, therefore, this power of the heart is feebly exerted, as during inaction, although it is sufficiently strong to carry the blood to the ultimate tissue, it is nevertheless not strong enough to carry it through it: at least, not to carry it through with the same rapidity with which it is brought to it. The result is evident: the ultimate tissue being thus filled faster than it is emptied, there must accrue accumulation; that is, congestion in the delicate and most important vessels which compose this tissue; and also in the larger veins, whose office it is to convey the blood from this tissue to the heart. For if the propelling power of the heart be but feebly exerted within the tissue itself, it must be still more feebly exerted on vessels which are situated beyond it, and, consequently, farther removed from the propelling force.

One of the chief conditions of the body, therefore, in that general ill state of health usually denominated "indigestion," "bilious disorder," "dyspepsia," is congestion of blood in the ultimate tissue of our organs—the brain, the lungs, the spinal marrow, the stomach, the ganglionic system, the liver, bowels, &c.—all the organs concerned in the nutrition of the body, but especially of the arterial and venous capillaries of the nerves and nervous centres. There is congestion of the brain: the veins of this organ do not carry away the black deteriorated blood with sufficient expedition; they (the veins) become distended; and thus, occupying more room than they ought to do, exert a very considerable degree of pressure upon the surrounding parts—the origin of nerves, &c. But, besides the great evil resulting from this general pressure on, and this great irritation set up within, the brain and nerves (like the irritation produced by a blood-shotten eye), there is another evil produced, by the accumulation of venous blood in the brain, equally important. For it is a well-ascertained fact, that this black venous blood has a direct influence in diminishing contractility and sensibility—it is even capable of utterly destroying them; it is at open war with life—it exercises a destructive and paralyzing influence on the living powers, and, wherever it accumulates, poisons the life of the part. Comparing the human machine to a watch, and contractility and sensibility to the elasticity of the main-spring, upon which the motions of the watch depend, then, I say, venous blood has a positive and direct influence in weakening this elasticity; and the several actions constituting life are injured by congestion of venous blood, as the movements of a watch would be injured by any cause which weakened the elasticity of its main-spring. The watch would lose time—its movements would be too feeble—it would go too slow. So with the human machine: its actions are enfeebled—its nutrition is languidly and imperfectly performed—it goes too slow. The indication of time is the final cause of the movement of the watch; and the nutrition of the body is the final cause of the nutritive actions. Both these final causes depend upon the healthy existence of the first cause—the elasticity of the main-spring in the watch, and contractility and sensibility in the body. It is manifest, therefore, that whatever weakens the energy of the first cause must have a direct tendency to interfere with, and prevent the fulfilment of, the final cause: that is, in the human machine, the nutrition of the body. It is this destructive influence of venous blood on the life of our organs which has caused me purposely to speak of it with so much bitterness, whenever I have had occasion to mention it. I did so in order that your mind might be thoroughly impressed with a sense of the great and necessary distinction to be drawn between venous and arterial blood. Congestion in the *venous* capillaries of the nervous system produces torpor, lassitude, and oppression of spirits. Congestion in the *arterial* capillaries produces preternatural excitability, restlessness, irritability of temper, headache, heat on the top of the head, and all those symptoms classed under the head of irritation, whether of mind or body.

It is difficult to make this impression on the mind of a person not conversant with physiology. Both fluids go by the general name of blood ; and this increases the difficulty, weakens the distinction, and produces confusion in the mind. It seems difficult to be understood how blood can, at once, be the support of life, and yet destructive to it. Blood is blood :—and people generally are not aware that there are two sorts of blood in the body. Blood is blood :—true ; but arterial blood is not venous blood : and there is not more difference between champagne and ditch-water, than between these two kinds of blood.

One of the most common symptoms of the disordered condition of the body, now under consideration, is the appearance of small, black, irregular-shaped specks, resembling little pieces of broken cobweb, floating before the eyes. This arises from congestion of venous blood in and about the nerve of vision—the optic nerve—and retina. This nerve is compressed by the gorged veins entering into its own structure, and also of the veins immediately surrounding it. The energies of the nerve are partially paralyzed by the debilitating and devitalizing influence of the nervous blood which has accumulated within and around it. This nerve, too, like every other part of the body, is in a state of perpetual disorganization and reparation. The disorganization goes on as usual ; but the reparation (that is, its nutrition) does *not* go on as usual : its nutrition is imperfectly performed. The capillary arteries entering into the structure of its elementary tissue are filled with a blood, from which the nutritious particles, necessary to repair the waste which it perpetually undergoes, cannot be separated in sufficient abundance : its structure, therefore, becomes flaccid, and its energies consequently enfeebled. If this state of the nerve go on increasing—if these little black spots go on multiplying and increasing in size, until they form but one black speck—that is, complete darkness—you then have at once the disease called *amaurosis*, "blindness," resulting from palsy of the optic nerve and retina.

Imperfect hearing, and ringing in the ears, are also very common symptoms of indigestion. These arise from a condition of the nerves of hearing—the auditory nerves—similar to that which I have just described as incidental to the nerve of vision. All the nerves of the body, with the questionable exception of one, arise from the brain and spinal marrow; and all are liable to be thus injuriously affected by venous congestion in the elementary tissue of these two organs. It is to this condition of the nervous system that are to be attributed all those oppressive feelings of lassitude, ennui, mental imbecility, &c., so prominently characteristic of the hypochondriacal dyspeptic.

Then we have, too, congestion in the lungs, interfering with those important changes which should be effected on the blood in those viscera.

Then we have congestion in the stomach. In this viscus the food is destined to undergo the first important change towards final assimilation; that is, nutrition. This change is effected by admixture with the gastric juice. The gastric juice is secreted, that is, separated from the blood—that is, formed or manufactured, as it were, by the arteries entering into the composition of the elementary structure of that organ. But, in order that this juice may be secreted in sufficient quantity, it is necessary that the elementary tissue of its blood-vessels should be plentifully supplied with pure arterial blood: whereas, in congestion, this tissue, as I have before shown, is gorged with venous blood. The necessary quantity of gastric juice, therefore, cannot be formed; and that portion which is secreted is not of a healthy sort. The direct result of all this is, that the very first necessary change which should be wrought upon the food, in order that it may nourish our bodies, is very imperfectly effected—the chyme is of unsound quality. The next result is this:—the chyme, by admixture with certain other juices which it meets with in the bowels, is destined to become chyle. But the chyme, being of vicious quality, the chyle which is formed from it must be also vicious; at all events, it must be deficient in quantity. Certainly, it is impossible to suppose that as much perfect chyle can be elaborated out of bad chyme as out of good: you might as well hope to make as much good butter out of bad cream, or out of cream and water, as out of pure cream. The chyle, therefore, is deficient in quantity: but this chyle is destined to become blood. The chyle, therefore, being deficient, the blood resulting from it must also be deficient. But the blood is, in fact, as I have shown you before, the real food on which the body feeds, by which it is nourished, and its strength supported; and this food being scantily supplied, the strength, of course, is but ill supported.

I have said that the chyme is converted into chyle by admixture with certain juices which it meets with in the bowels. But the same causes which we have seen producing a deficiency of gastric juice, produce also a deficiency in those juices, by commixture with which the chyme is converted into chyle. Here, then, is another cause which tends to diminish the quantity of chyle; and, consequently, blood—the nourishment which we ought to derive from our food.

But there is yet another mischievous result accruing from a congested condition of the stomach and bowels, besides that of deficient and unhealthy gastric juice. In that condition of the health which I am endeavoring to describe, the stomach and bowels actually secrete *air*. It is a thoroughly-established fact, that air, wind, flatus, is actually formed from the blood, and poured into the stomach and bowels by those arteries which ought to form only gastric juice. Now this wind not only does no good in the stomach and bowels, but it does a vast deal of harm: for, besides the evil effects which it produces by its pernicious qualities, it violently distends these organs, stretching, and separating, and thus greatly weakening and destroying, the firmness and compactness of their ultimate tissue.

To give you a still further and clearer idea of the manner in which the secretions of the body may be altered in their quality, as well as diminished in quantity, I have only to direct your attention to a foul tongue. Look at the tongue of a sick man: instead of being bathed in the natural, pellucid, and fluid secretion of the mouth (saliva), it is covered with a thick, and filthy, and pasty crust, which is actually solid. Is it possible to conceive that the offices intended to be fulfilled by the saliva can be properly effected, or effected at all, by the nasty, pasty filth which you behold in his mouth, instead of saliva? This filth is secreted from the blood; and poured into the mouth by those arteries which ought to secrete only that thin, clear, pellucid fluid called saliva. Now this is an example of vitiated secretion which you can actually see, and which cannot, therefore, be doubted or questioned. Can you have any difficulty in believing that the other secretions of the body, which are all formed from the blood, may, in like manner, be equally altered and vitiated? Can you, moreover, have any difficulty in believing that the body must, of necessity, be badly nourished in this state of the secretions, seeing that it is by the agency of these very secretions that our food is converted into blood, which blood is the sole source from which our bodies derive nutrition and support ?

I have now described to you what I believe to be the actual condition of the body in that sickly state of the health usually denominated "indigestion;" a condition from which few persons in the upper and middling ranks of life are wholly and perfectly free. And you will observe, that all the mischief arises primarily from depressed contractility and excessive sensibility, which are themselves the immediate offspring of excessive refinement.

E. JOHNSON.

LETTER IX.

Umberslade Hall, near Birmingham.

MY DEAR JOHN,

In my last Letter, I stated to you, that I believe sanguineous congestion, in the ultimate tissue of our organs, especially of the brain and other parts of the nervous system, constitutes that morbid and multiform disease usually denominated "indigestion," or "dyspepsia."

The immediate cause of this congestion I believe to be a sleepy, feeble, and inefficient circulation, occasioned by the peculiar habits of artificial society, the direct effect of which is to impair the contractile vigor of our organs.

Indeed, when one considers the amazing exertions which the human body is manifestly constructed for the purpose of undergoing—when one sees, every day, the extraordinary powers and wonderful activity which it is capable of exerting—and then, when one reflects upon the comparative physical inaction in which the lives of those are passed who are the victims of this disease—I mean the upper and middle orders, especially of females, and such of the lower whose occupations are sedentary—when all this is considered, I say, one is astonished, not that the health of the machine should suffer, but that it should continue to exist at all. It seems really wonderful that organs of such elaborate and delicate workmanship should be able to perform their functions at all, under circumstances so diametrically opposite as those of action and inaction. Which of these two conditions, however, is the better suited to the body, daily and hourly experience shows; since robust health, and great physical strength, are almost only to be met with in the ranks of those who earn their livelihood by bodily exertion; and since that sickly habit of body, concerning which I am speaking, is almost solely incident to those whose lives are physically inactive.

Who ever heard of a bilious post-boy, or dyspeptic ploughman? It is not amongst carpenters, and bricklayers, and sawyers, and agricultural laborers, that you will meet with the dyspeptic ; but in the halls and saloons of the great, the dusky counting-houses and gas-illumined shops of the trader, and in the ghost-like and dwarfish ranks of the pale and spectral silk-weaver. Indeed, of the many hundreds of those who have come under my observation during thirteen years, I never remember to have seen a single silk-weaver who was not more or less dyspeptic.

Another important cause of languid and inefficient circulation is, the manner in which we surround ourselves with what are called comforts. We clothe ourselves in flannel, and envelop ourselves in great coats abroad ; and when at home, we close the doors, let down the window-curtains, draw a chair to the fire, bury our feet in the wool of the hearth-rug, and make our servants wear slippers that they may not disturb us.

Now, these same comforts exert an opiate influence on the system—an influence directly lulling and somniferous. I surely shall not be called upon to prove this. Who has not himself experienced that almost irresistible disposition to sleep, which an easy chair, a warm room, a good fire, and silence, induce? And who will not sleep more soundly in a darkened room, on a down-bed, surrounded by curtain drapery, and well covered with blankets, than on a straw mattress, scantily covered, uncurtained, in a garret?

Those, therefore, who surround themselves with these seductive "comforts," place themselves precisely in the situation of opium-eaters. They submit their bodies to the same influence, and suffer the same evils, although the cause be different. "Comforts" are opiates—anodynes—narcotics ; as certainly so as opium itself, although not in so powerful a degree. They lower the tone of the nervous system, and lessen the contractile energy with which the circulation is carried on. The lover of "comforts," therefore, must neither censure nor ridicule the eater of opium : he is himself guilty of the same fault, and will certainly reap the same harvest. I say, their fault is the same: they both are producing the same effect, only by different means: they are both travelling to the same point, only by different roads.

Like hemlock, then—like the deadly nightshade—like opium, and other poisonous narcotics—"comforts," as we are pleased to term them, but whose proper name is "*luxury*," have the direct effect of lowering the tone and lessening the activity of the living actions ; and of inducing that condition of the body called lassitude, excessive sensibility, &c.

Light, and wet, and wind, and cold, and noise, &c., &c., are what are enumerated among the discomforts of life. But these, and the like of these, are the natural whips and spurs which keep the living actions, as it were, awake; they form a part of man's natural condition: they form a part of the means which nature has contrived, to keep up the activity of the machine—to prevent its going to sleep, like a lazy horse, when he no longer hears the whip, or feels the spur. These discomforts, as they are called, are to be considered as so many incentives to exertion; for by exertion they not only (at least many of them) cease to be discomforts, but become real pleasures. What, for instance, can be more delicious than the bright and frosty freshness of the air to the active skater? What more luxurious than water to the athletic swimmer?

These discomforts are component parts of the system of this world: and man was made, and expressly fitted, to inhabit this world. In his construction, nature intended that his system should be adapted to the system of the world, and not that the system of the world should be altered in order to be adapted to his own sensual and animal gratification—his own love of luxury. Yet this is what we are perpetually laboring to do, in surrounding ourselves with these same comforts: for every comfort is, in fact, no more than the absence of some supposed discomfort. But, as I have shown, these discomforts form a component part of the general system of the world: and, therefore, to get rid of them, is to alter that system from its original order. But as the system of man was adapted to the system of the world at the creation, it follows, that to alter the system, or rather circumstances, of the world posteriorly to the creation, is, in fact, to destroy the adaptation then made and effected by the Author of our existence—to destroy the relation then instituted between ourselves and the things and circumstances wherewith we are surrounded. But thus it is:—instead of being satisfied with Nature's adaptation of our system to the world, we seek to alter the world, and the order and circumstances of things, in order to adapt them to our system. Thus the sweet breath of heaven is carefully excluded by windows, and shutters, and curtains: and the cold most assiduously dispelled by fire and flannel. The rain must not wet us; the wind must not blow upon us; cold must not approach us. Thus we surround ourselves with new circumstances, in the place of those to live among which we were expressly constructed and contrived.

Nature always husbands her means, and ever produces the greatest possible number of effects from the fewest possible causes. Accordingly, seeing that the system of man was destined to inhabit the world, she seized upon certain parts of the system of that world, and made them subservient to the existence of her new creation. Thus, air is absolutely necessary to the existence of man. She might have constructed him so as to live without air; but then some other contrivance must have been adopted; and to have instituted a contrivance which did not exist, in order to effect a purpose that might be as well effected by a contrivance already existing (viz. air), would have been to waste her means, and unnecessarily exhaust her energies, which she never does. And as air, which is one of the component parts of the system of the world, is absolutely necessary to the existence of man, so the other so-called "discomforts" of life, such as cold, wet, hard fare, hard lodging, which are also component parts of the system of the world, are absolutely necessary to the perfection of his *healthy* existence.

As in the case of air, so in the case of other discomforts of life, Nature, it is true, could have fulfilled her task without them: she might have contrived other means to preserve the health of the human machine: but these were ready-made to her hand; and she made use of them at once, as she always does, rather than waste her energies by the invention of new ones.

Thus are all the systems of things, animate and inanimate, dovetailed into each other. One supports another, and is itself by others supported. This is the invariable conduct of Nature. If she had to prevent two houses from falling, she would not get a prop for this and a prop for that;—no; she would make one house prop up the other.

It is this propping and dovetailing of one system with another which constitutes what I mean by the relation of one system to another; as, for instance, our own system to that of the world; and which makes it so impossible to destroy, or in any way interfere with, that relation or adaptation, without mischief to the individual system; (which is thus, as it were, withdrawn from the support of the rest) without injury to the beauty and harmony of the whole.

Now, the things which we are accustomed to consider "discomforts" are the very contrivances by means of which our system is interwoven, as it were, with and into the system of the world which we inhabit, and are absolutely necessary to secure this connection. I will explain this.

You will, I hope, remember, that there are four conditions absolutely necessary to the existence of all animals; viz., organism, contractility, sensibility, and *stimuli*. The principal of these stimuli is the blood: but this is by no means the only one. There are many others; such as light, heat, electricity, and the excitement produced through the medium of our organs of hearing and seeing, &c.: but, besides these, there are also others. And what are these others, my dear John? Why, precisely the very circumstances of our natural existence which are now under discussion:—I mean, the very self-same "discomforts" aforesaid. They form a part of the necessary and natural stimuli. As "comfort" (that is, the absence of all "discomfort") has the effect of lulling the system to lassitude and sloth; so "discomfort," which is the opposite of "comfort," produces an opposite effect; viz., that of rousing the system to energy and action. He who sleeps on the hill-side, unsheltered, is not likely to *sleep* too long.

It was ordained that the human heart should continue to pulsate for a certain number of years. In order to accomplish this, it was necessary to afford to it a perpetual supply of stimulus, to a given amount. If the blood alone were capable of supplying this necessary given amount, then, when the being to whom this heart belonged came to be placed in the world which he was destined to inhabit, and within the operation of these other stimuli, he would immediately suffer by excessive stimulation, being sufficiently stimulated by the blood before he became submitted to the action of these additional stimuli. But foreseeing this evil, Nature has so ordered it, that the stimulating properties of the blood are alone insufficient; and this insufficiency of stimulation is made up to the necessary amount by the adventitious stimuli afforded by the nature of the circumstances with which he is surrounded, and which he is pleased to denominate "discomforts." To remove these circumstances, therefore, is to remove a certain number of the stimuli which are absolutely necessary to the healthy activity of the living actions.

You will now clearly understand what I meant, when I said that our so-called "discomforts" are the necessary whips and spurs which keep the living energies awake. You will also now see how it is, that what we call "comforts" operate upon us like opiates; —since, to acquire a "comfort" is only to remove a "discomfort;" and to remove what keeps us awake, is the same thing as to administer what will send us to sleep.

The indulgences, therefore, wherewith even young and healthy men indulge themselves—the "comforts," as they call them, of flannel, warm clothing, closed doors, carpeted rooms, soft beds, hot food, are infinitely worse than absurd: because the opposites of all these luxuries, so far from being injurious to health, are absolutely necessary to it. We actually kill ourselves with "comforts." It is absolutely disgusting to see the excessive care and caution with which great fellows, with rough beards on their chins, and with fists large and strong enough to fell an ox, and legs long enough to bestride the Thames—I say, it is neither more nor less than disgusting to see these lackadaisical women in the likeness of men, or, rather, these monsters, which are neither men nor women —I say, it is literally disgusting, and degrading too—degrading to our nature, to our being—degrading to the physical energies of Nature's master-work—to see the care and pains-taking with which these abortive monstrosities, the progeny of a morbid and excessive refinement, protect their delicate and precious persons from a few drops of rain, or a little mist, or a little unusual inclemency of the weather, of whatever kind.

I got into a coach, a mile from London, the other day, because there was no room outside. The weather was dry, but cold and sharp. In the corner of the coach there sat a mighty combination of bone and muscle, thew and sinew, all assisting in the formation of what should have been a man. He was, at least, six feet high, and "bearded like the pard;" and seemed as well able to carry the coach as the coach was to carry him. As soon as I entered the vehicle, I let down the window; but before I had quite succeeded in doing so, there issued, from amidst the cloaks, and coats, and shawls, and wrappings, and mufflings, in which this great thing had enveloped itself, a voice of supplication and woe: "For God's sake, do not let the window down! I am so susceptible—so extremely susceptible!"

Look at the delicate and fragile plant in your garden! see how it is buffeted by the wind, and alternately scorched by the sun, and deluged by the rain, and frozen by the frost, and spattered by the mud, and brushed and bruised by the passenger's foot! Yet how greenly and healthily it grows! Take it into your parlor, and warm it by the fire, and curtain it with flannel, and defend it from the cold, and the wind, and the rain, and the rude contact of the traveller's foot, and the other "discomforts" of its out-of-door existence.—What think you? will it continue to flourish as greenly and healthily as before? "Oh! but," say you, "there is a difference between a man and a cabbage!" A difference! Why I know there are many differences! A man does not bear leaves, and look green; a cabbage has neither arms nor legs; and though it has as good a heart as many who rejoice in the name and nature of man, still that heart contains no blood. But what of all this? To constitute an analogy, it is not necessary that there should be agreement in every particular. At this rate there would be no analogy between man and woman, nor even between man and man; for there are, probably, no two men in existence exactly alike. But, in all that concerns our present purpose, in all essentialities, the man and the plant are perfectly analogous: they are both living beings, destined to exist under certain circumstances—living systems, destined to occupy a certain position within the circumference of that circle of existence which constitutes the universal whole. We have seen, and we know, that we cannot remove the one (that is, the plant) from its prescribed position, without great injury to its health; why, then, do we presume that we may, nevertheless, remove the other with impunity? Those who are not conversant with animal and vegetable physiology will be astonished, upon examination of the subject, to find how little indeed is the real and essential difference between plants and animals. In all, life is the same; more or less complex, but still the same; consisting, in all, of a number of effects, resulting from, and depending upon, the four grand conditions of matter before mentioned; viz., *organism*, *contractility*, *sensibility*, and *stimuli*.

To show that the "discomforts" of life, or hardships, as they are called, have no influence in producing disease, but, on the contrary, serve only to harden the system against it, Dr. J. Johnson has most aptly quoted some remarkable historical illustrations. "One of the earliest and most memorable illustrations," says he, "will be found in the celebrated retreat of the *ten thousand Greeks*, under Xenophon and Cheirisophus, after the fall of Cyrus on the plains of Cunaxa. This band of auxiliaries were left without commanders, money, or provisions, to traverse a space of twelve hundred leagues, under constant alarms from the attacks of barbarous and successive swarms of enemies. They had to cross rapid rivers, penetrate gloomy forests, drag their weary way over vast and burning deserts, scale the summits of rugged mountains, and wade through deep snows and pestilent morasses, in continual danger of death—or capture, which was far worse than death. This retreat is nearly unparalleled in the annals of war, for difficulties and perils. During two hundred and fifteen days of almost uninterrupted and toilsome march, often in the face of the enemy, often between two enemies, and engaged in front and rear at the same moment, the army lost an uncertain, but not a great number of men: partly by the darts and arrows of the barbarians; partly by desertion; partly by drowning in rivers, or sinking in morasses; partly by perishing in the snows of the Armenian mountains;—but not *one by sickness!*" He mentions also, the case of Byron and his crew. "Although nine-tenths of the original crew appeared to have perished by drowning or starvation, Byron makes no mention of sickness, during any period of the long and unparalleled series of sufferings to which that ill-fated ship's company was doomed." The retreat of Sir John Moore through the mountains of Spain—the sufferings of the crew of the "Bounty," under Captain Bligh—and the retreat of the French from Moscow—are also quoted, in proof of the same principle.

Another prevalent cause of indigestion is the depressing influence of anxiety. In the present day, with men engaged in business, the mind is scarcely ever free from care: for business is not now, as formerly, a simple matter of buying and selling, and living by the profits: it is now, rather, a matter of speculative gaming. Every trader, almost, is a speculator; and his mind is, consequently, kept perpetually vibrating between hope and fear; for he knows and feels that the turning of a straw may make him, or mar him, forever. Never was the maxim, "Habe rem," &c., more religiously observed than in the present day. No man is satisfied to live and rear his family to tread in his own steps. Every man is striving to be wealthy. Men seem to have forgotten that the end of existence is happiness. They appear to have adopted the belief, that they were created for no other earthly purpose than the advancement of their condition in the ranks of society. They seem content to pass through life without enjoyment; to exist in any way, no matter how miserably and morbidly, so long as they can achieve this—apparently, to them, the sole object of their existence: thus utterly losing sight of the end, in the eagerness of their pursuit after the means.

Another cause of that degenerate state of health of which I have been speaking, is eating too much. All other animals eat because they are hungry, and drink because they are thirsty: man eats because it is dinner-time; and, having eaten enough, uses stimulating drinks, sweet puddings and piquant pastries, *variety of food,* in order to enable him to eat more; and then, feeling himself uncomfortably distended, drinks again, with a view to relieve the sense of oppression under which he finds himself laboring. Man, I believe, is the only animal who eats in order to induce himself to drink; and drinks in order to induce himself to eat. No other animal than himself requires any relish, saving only that of hunger and thirst.

I believe I have now enumerated what I consider as the principal causes of that disordered condition of the health, called "indigestion:" and you will observe, that they all arise from the artificial condition of society in which we live. In my next Letter I shall point out what I believe to be the only means of avoiding and remedying it.

I know it will be said in objection to what I have advanced against our luxurious modes and habits of life, that the Almighty has bestowed upon us a superior intellect for the purpose of enabling us to surround ourselves with these comforts (which are denied to the inferior animals), and in order that we might indulge in all *innocent* gratifications. Yes—but these habits are NOT innocent. How can they be innocent if (as they unquestionably and avowedly do) they injure our health, shorten our lives, and entail a sickly constitution on our offspring?

I am, dear John,
Yours ever,
E. JOHNSON.

LETTER X.

Umberslade Hall, near Birmingham.

MY DEAR JOHN,

As I have placed excessive eating among the cause of depreciated health, so must I now mention temperance in food as one of the prime remedies for it, and preventives against it. Learn, therefore, now, " *Quæ virtus et quanta sit vivere parvo.*"—What and how great is the advantage of living upon little.

When we consider that the manner in which life is supported is by a perpetual wasting of the body, and a perpetual reproduction of it out of the blood—and when we remember that the simple and sole object in eating is to make up to the blood the deficiency thus occasioned in it—it must be manifest to us, that the exact amount of food required daily is, precisely just so much as shall be sufficient to restore to the blood just as much as the blood has lost in supplying the waste which the body has undergone during twenty-four hours of life.

To make this more simple and clear, let us suppose, for argument's sake, that the waste of the body in twenty-four hours is just twenty-four ounces. Now, when these lost twenty-four ounces of the body have been restored to it out of the blood, then the blood will have lost twenty-four ounces: and the object of eating being wholly and exclusively to supply this deficiency thus produced in the blood, it is perfectly evident that the quantity of food required in twenty-four hours is precisely so much as shall be capable of conversion into twenty-four ounces of blood; that being the exact supposed quantity which the blood had lost in supplying the waste of the body in twenty-four hours.

I do not mean to say that twenty-four ounces do indeed form the precise quantity of daily waste; but it seemed necessary to fix on some definite and specified quantity, in order to illustrate more plainly the principle which ought to guide us in eating. There is, in fact, no fixed quantity of waste: for the quantity must always be in proportion to the extent of bodily exertion: and, for the same reason, the quantity of food daily necessary can neither be fixed, definite, nor specified.

Now, if a man eat more food than is necessary to supply the loss which the blood has suffered, one of these two things must happen :—it must either be assimilated, or not assimilated ; or, to use the common erroneous language, digested, or not digested.

If it be assimilated—that is, converted into blood—then it is clear that there will be more blood in the vessels than there ought to be. Let me illustrate again. Suppose the case of a healthy man—so healthy that he cannot be healthier. Let us suppose the whole quantity of blood in his body to be thirty pounds. Let us further suppose, that in twenty-four hours, one pound of his blood is lost in supplying the waste of the body. Now, if this man eat, in one day, so much food as will produce a pound and half of blood, what follows? Why, that his blood has lost a pound of its volume, and gained a pound and half in its stead: or, in other words, that the whole quantity of blood has been augmented by just half a pound ;—so that his system now contains just half a pound too much. If this man were to go on adding half a pound to his stock of blood—and if it were possible for him to escape apoplexy or some other deadly disease—and if Nature, foreseeing that her children would turn out to be gourmandizers, had not, in some measure, guarded against the evil—it is plain that his blood-vessels must soon actually burst under the distension. But nature has, though only in part, made a provision against this evil: for when, after having supplied the waste of the body, there is still remaining an undue quantity of blood in the vessels, the vessels relieve themselves, and reduce the quantity of blood, by the secretion of fat ; thus restoring the blood's volume to a due standard.

But the fat, *quasi* fat, is of no use whatever to the body: it does not add to its strength : on the contrary, it is an incumbrance to its machinery, and, in more ways than one, is an evil. The fat, *quasi* fat, does not form a necessary part of the body, any more than the padding and wadding of a fashionable coat form a necessary part of the coat. The padding of the coat does not add an iota to the strength and quality of the original texture of the cloth ; and the coat would be just as good without it. All that the padding does, is to add to the beauty of its appearance. So of the fat: it contributes nothing to the health and strength of the body ; nor does it form a necessary part of the body : it might be all cut away without detriment to the body, and even, if it were not for the skin which covers it, almost without pain: it has nothing whatever to do with the body, saving only as it adds to the beauty of its symmetrical proportion—it has no more concern with the health and strength of the body, than the coat-padding has with the texture of the cloth whereof the coat is made.

I know that in the leanest persons there is still a certain portion of fat deposited in particular parts, as behind the eye, &c.; but this is merely for the purpose of giving to the *tout ensemble* of the body a certain appearance of symmetry and beauty of outline. What, for instance, has the fat behind the eye to do with the power of seeing? But, without it, the eye would have the disagreeable and sinister appearance of being sunken too deeply in the head.

He, therefore, who eats too much, even though he assimilates what he eats, and should be fortunate enough to escape apoplexy and some other deadly disease, does not add a single iota's worth to his strength: he only accumulates fat, and incurs the evils thereunto appertaining—one amongst the many of which I will mention—I mean, the accumulation of fat about the heart—making him puff and blow like a grampus, and interfering, to a most dangerous degree, with the heart's action.

But neither does he add to the size and weight of his body, properly so called. He may indeed add to the size and weight of his body's fatty envelope, as a tailor may add to the padding of the coat; but both the one and the other, properly so called, still remain unaltered.

A man's strength resides in his arterial current—in his muscles, and bones, and tendons, and ligaments—in his brawn and sinew; and his degree of strength depends upon the vigor, size, and substance of these; and if he were to eat a hecatomb of oxen every morning for his breakfast, and, like Gargantua, swallow a windmill for his dinner and a church for his supper, he could not add to their size and substance one atom, nor alter their original healthy dimensions—no, not in the estimation of a single hair.

Remember, then, my dear John, that it is a most miserable and mischievous fallacy to suppose that the more a man eats, the stronger he grows. If a man require daily one pound of food to supply his diurnal waste, recollect, that although he may eat double that quantity, yet he will be not one atom stronger, nor longer, nor broader, than if he had eaten no more than the one necessary pound. He may have enveloped himself in an extra layer of fat—he may have added another portion of padding to the coat; but he himself, like the coat, will remain in *statu quo*, with the chance of being found, some morning, dead of an apoplexy. He who eats more than he wastes, with the view of making himself stronger, is guilty of precisely the same folly as he who should continue to pour water into a vessel which is already full, with the view of filling it fuller.

But, in some constitutions, if a man eat greatly too much, the secretion of fat may not be sufficient to relieve the overburthened vessels. Now, if this man should escape the usual disease resulting from plethora, then there is, in literal fact, a very great danger that some one or other of his vessels may, indeed, actually burst; and so destroy him, by bleeding from the lungs, or some other active and deadly hæmorrhage. Tell me, John, what warranty have you that your constitution is not one of this kind?

We arrive, therefore, at this inevitable conclusion, viz., that he who eats more than is necessary to supply his waste, even although the whole be well and truly assimilated, not only does not increase his strength thereby, but really incurs the danger of destruction from several probable causes, and is constantly walking heedlessly in the "valley of the shadow of death."

But, if the other and more frequent circumstance happen—if what is eaten be not properly assimilated—then that which remains unassimilated becomes a source of great irritation and numerous morbid symptoms, as I have explained to you in a former letter: it ferments in the stomach and bowels, as it would do in any other close, warm place; and the gases given out during this fermentation, and the acids generated thereby, are neither more nor less than poisons, and, of course, highly injurious to health.

If, therefore, a man, under these circumstances, eat more than is necessary, nothing can be more manifest than that he only adds to the evil he wishes to remove. For, since his assimilating powers can only assimilate just sufficient to supply the body's waste—and, in these circumstances, not even so much—it is surely most clearly evident, that by adding to the quantity eaten, he only adds to the quantity which is destined to be left unassimilated, and, therefore, to give out a still greater portion of those poisonous gases and acids above-mentioned;—and an increased quantity of these poisons must produce an increased quantity of mischief to the health. And thus it becomes plain, that so far from growing stronger, he will only become weaker, and worse nourished, the more he eats.

Thus, from the very nature of the animal system—from the very manner in which life is supported—it is manifestly impossible to add to the natural standard of health and strength by increasing our quantity of food—whether that food be well assimilated or not: and it is equally clear, that when the health is weak, and the assimilating powers thereof feeble, that eating more is not the proper remedy. For, certainly, the assimilating powers, which are not equal to the assimilation of one pound of food, must be still more unequal to the assimilation of two. And it is also plain, that, under these circumstances, the proper way to improve the health is to diminish the amount of daily food; since those powers, which are inadequate to the assimilation of a pound, may, nevertheless, be equal to the assimilation of eight ounces.

I have said, that the quantity of food taken daily should be just sufficient to restore to the blood what the blood has lost in restoring the waste of the body, and that it should always be proportioned to the degree of bodily exertion undergone.

You might here very properly inquire, how we are to know the exact amount of this daily waste, so as to apportion the quantity of food thereto. Are we to weigh ourselves every morning, in order to ascertain this important fact? No, my dear John: Nature has not left any part of her master-miracle incomplete; which it would have been, assuredly, had she not provided us with infallible means of knowing, not only when, but how much, we ought to eat and drink.

When you are excessively thirsty, and when you are in the act of quenching your thirst with a draught of cold water (which I shrewdly suspect is but seldom), tell me, how do you know when you have drunk enough? One token, by which you know this, is the cessation of thirst;—and this, of itself, should be sufficient;—and, in truth, so it is, when you drink water, I dare say. But there is still another; and one which not only informs you when you have drunk enough, but which also prevents you from drinking more. While you are in the act of drinking, and before your thirst has been allayed, how rich, how sweet, how delicious is the draught, though it be but water! But no sooner has your thirst been quenched, than, behold, in an instant, all its sweetness, all its deliciousness has vanished! In a moment, how insipid it has become! It is now distasteful to the palate—positively disagreeable—it has lost its relish. To him then who requires drink, water is delicious:—for him who does not require drink, water has not only no relish, but impresses the palate disagreeably, by its very insipidity. Carry this a step farther. To a man laboring under the very last degree of thirst, even foul ditch-water would be a delicious draught; but his thirst having been quenched, he would turn from it with disgust. In this instance of water-drinking, then, it is clear that the relish depends, not on any flavor residing in the water, but on a certain condition of the body. If, therefore, we only took drink when drink was required, pure water would be sufficiently delicious: but we seek to give to our drink certain exciting and racy flavors, as a substitute for that relish which should, of right, reside in ourselves;—and we do this in order to enable ourselves to drink when drink is not required. It is absurd, therefore, to say that you cannot drink water because you do not like it, for this only proves that you do not want it; since the relish with which you enjoy drink depends upon the fact of your requiring drink, and not at all upon the nature of the drink itself.

Now, apply all this to eating, instead of drinking. Place before a hungry ploughman stale bread and fat pork, flanked by a jug of cold water. While his hunger remains unappeased, he will eat and drink with an eager relish: but when his hunger has been satisfied, the bread and meat and water will at once have lost what he before supposed to be their delicious flavor. I say "supposed;" because, in fact, the relish only existed in his own appetite—in the condition of the nerves of the palate, produced by hunger. And it is to produce artificially, and when it is not required, this condition of the palatal nerves that we use highly-flavored food; for, in eating, we seem to have entirely lost sight of the true object of food; and only eat for the sake of the enjoyment which the act of eating affords us.—But to return to our ploughman.

When his appetite has been fully appeased, his food seems to have lost at once all its flavor: the attempt to eat more would now produce a feeling of disgust; and, if he were to persist, would, in all probability, make him sick.

If, then, we ate only simple and natural food, plainly cooked, there would be no danger of eating too much;—the loss of relish, and the feeling of disgust consequent upon satisfied hunger, would make it impossible. And I affirm, that there is just as much reason to believe that this sense of disgust is as much, and as truly, a natural token, intended to warn us that we have eaten enough, as the sense of hunger is a token that we require food.

Hunger is an instinct;—disgust is an instinct. Instinct signifies an inward pricking, an internal sensation, prompting us to some external action. It is by virtue of this, that the infant is enabled, untaught, to perform the complicated action of sucking. Nature has supplied us liberally with these instincts—instincts teaching us, not only what to do, but also what to leave undone. These warning sensations may be called Nature's code of instinctive laws for the regulation of man's conduct, as it regards the preservation of his health. Thus, hunger teaches us when to eat:—thirst, when to drink:—and disgust or disrelish, when we have eaten and drunken enough. Weariness teaches us when to rest: and that feeling (to which I can give no name) which induces the healthy child to run, and leap, and toss its arms, and shout—which causes the horse in his meadow to curvet and capriole, and exult in his strength—it is this feeling, call it what you please, which teaches us that we have rested enough, and that the time for action has come. Drowsiness teaches us that we require sleep: the internal sensation, whatever it be, which awakens us, teaches us that we have slept enough. But I need not multiply instances. The voice of Nature is, in fact, never silent;—for when we are doing what she requires, in obedience to her laws, and when, therefore, it is not necessary to warn us, even then her encouraging voice is heard in the pleasure which we feel.

In the infancy of creation—

"When the world was in its prime;
When the fresh stars had just begun
Their race of glory; and young Time
Told his first birthdays by the sun;"—

while man was yet content to listen with respect to the lessons of his parent Nature—he regulated his conduct solely by these instinctive laws. But Refinement, with her harlot smile and syren voice, stole upon his retirement, and he no longer heeded the plain lessons of his simple teacher. The Appetites and Passions usurped her throne; and incontinently set themselves to work, to alter, amend, and modify her laws. But, unfortunately, they were all so drunk when they undertook the task, that they spilled the ink over the page, blotted the manuscript, and rendered it nearly illegible forever.

To illustrate this:—I have said, that as hunger instructs us when to eat, so disrelish teaches us when we should desist. But by what labor, and pains, and contrivances, has the unnatural art of cookery endeavored to annul this law? For what are the spices, and sauces, and gravies, and kickshaws of the cook, but so many provocatives to induce him to eat more who has already eaten enough?—to provoke him to drink who is not athirst—and him to eat who is not hungry? The very *ne-plus-ultra* of the cook's art is to destroy this sensation of disrelish; which is almost as necessary to our health as hunger itself. According to Dr. Fordyce, "it is a universal maxim" in the Black Art—that is, the art of cookery—"never to employ one spice, if more can be procured." Now, pray open both your eyes, and mark the object of this;—"the object in this case," says he, "being, to make the stomach bear a large quantity of food without nausea!" So that the object of modern cookery is, to cram into the stomach as much as it can possibly hold, without being sick. *Proh pudor!* Said I not well, when I called modern cookery the "Black Art?" Yet this is one of the elegances of modern refinement! We stimulate our palates with wine, that we may relish more food; and then swallow more food, that we may relish more wine:—

> "We swallow firebrands in place of food;
> And daggers of Crete are served us for confections."

And this is feeding according to the improved method—according to the rules and regulations of refined society! Why, the very hog that revels in the garbage of the shambles—all hog, and beast, obscene and filthy as he is—is, nevertheless, not beast enough for this! What difference does it make, in the true spirit and very reality of the thing—what real difference, I say, does it make—whether you force down your throat more food than you want, by means of a glass of wine, or by means of a long stick, as they cram Norfolk turkeys? "The rose by any other name would smell as sweet;" and cramming is cramming, call it by what name you please, and effect it how you will.

But it may be said, that if it were not for these provocatives, persons with delicate stomachs would not be able to eat at all. Nonsense. He who says this, is either a fool or a Jesuit. If they do not eat, it is because they have no appetite. What they want, then, is not food, but an appetite for food. They want one thing, but seek another. The stomach "asks for an egg, and they give it a stone." Let them use the necessary and natural means to procure an appetite, and they will require no other provocative: but they are "corrupti judices," and "malè verum examinant."

> ——"Leporem sectatus, equove
> Lassus ab indomito; vel si Romana fatigat
> Militia assuetum græcari; seu pila velox,
> Molliter austerum studio fallente laborem,
> Seu te discus agit; pete cedentem aera disco.
> Cum labor extuderit fastidia, siccus, inanis,
> Sperne cibum vilem; * * *
> * * * * cum sale panis
> Latrantem stomachum bene leniet.
> * * * * * * *
> * * * Tu pulmentaria quæra
> SUDANDO."*

Let me say a word or two on the subject of hunger. In the upper and middle ranks of life, I believe that true, genuine, honest, schoolboy hunger is almost wholly unknown. Is this because hunger is a feeling not proper to men as well as boys? Ask the shipwrecked sailor. No: it is because here also, as in the instance of disrelish before mentioned, modern habits have stepped in, and amended—should I not rather say mongrelized?—the natural feeling. It is true, that when

> "The tocsin of the soul—the dinner-bell"—

calls us to dinner, we feel a something which we call hunger: but this is not hunger; it is a sense of want, of the same nature as that which the dram-drinker feels when the hour for his dram comes round. It is the customary excitement which we miss and want: it is this, and not food, which the stomach is then craving. There is not one in a score, of those of whom I speak, who, when the tocsin sounds, although he may complain that he wants his dinner, could sit down with no other drink than water, and dine on bread and cold meat. Yet, surely! surely! bread and cold meat are all that genuine and natural hunger should require! What would you say to the beg-

* Translation. "After having wearied yourself with hunting, or backing an unbroken colt—or, if the Roman military exercises be too fatiguing for the effeminacy of your Greek habits—if either the tennis-ball or quoits please you better, seek the quoit-ground or tennis-court—and when exercise shall have banished fastidiousness, then, hungry and parched with thirst, despise common fare if you will. Bread and salt is fully sufficient to soothe a barking stomach. Seek thou no sauce other than SWEATING."

gar at your door, who should tell you that his stomach was so delicate that he could not eat cold meat and bread?

But if they could get it down, it would not allay the feeling which they call hunger. Why? Because that feeling is, in truth, not hunger, but a feeling which a pint of wine would allay more readily than food. Thus we eat for the sake of the stimulus which our highly-dressed dinners afford us; seeming to forget entirely that nourishing the body has anything to do with the matter. But to return.

The rule, therefore, which is to regulate your quantity of food, is to be found in that sensation of disrelish which invariably succeeds to satisfied appetite; provided always that your food be plain, and your drink water. If you be content to live thus, you will never eat too much, but you will always eat enough. But if you would rather incur the penalty of disease than forego the pleasure of dining daintily, all I can say is, you are welcome to do so: —but do not plead ignorance: blame only yourself.

One of the means, therefore, of preserving the health, is a spare diet;—I say "spare," because the upper and middle classes, together with that numerous class of persons consisting of manufacturers, whose employment is sedentary, such as weavers, tailors, shoemakers, milliners, &c., &c., with counting-house clerks, and journeymen tradesmen of the better order, such as mercers, linen drapers; and, indeed, shopkeepers of all grades, whose chief work consists in chaffering behind a counter;—I say "spare diet," because these persons undergo but little bodily labor, and the bodily waste is, consequently, small: they require, therefore, a correspondent small quantity of food; and if they be not careful to distinguish between true hunger and that feeling of want and languor which árises solely from a distressed state of the nervous system, resulting from the nature of their employment, and from their "cabin'd, cribbed, confined," and sedentary habits, they will be constantly falling into the error of eating too often and too much: because, mistaking this feeling for hunger, they will eat with a view to relieve it;—and for a short, a very short time, the stimulus afforded by the presence of food will relieve it. But if they do this, they will commit the grievous error of perpetually adding to the mischief which they seek to remedy: for this distressed state of the nervous system is peculiarly unfavorable to assimilation; and if they eat too often, or too much, they will inevitably become miserable dyspeptics. What they want is, excitement, not food. And how is this excitement to be procured? and in what should it consist? Be patient, my dear John; I will tell you presently.

STIMULANTS.

Are stimulants—by which I mean ardent spirits, wines, and strong ales—are stimulants necessary? Are they pernicious? or are they neither the one nor the other? I assert that they are, in every instance, as articles of diet, pernicious; and even as medicines, unnecessary; since we possess drugs which will answer the same intentions, in, at least, an equal degree. But it is only as articles of diet that we are here to consider them.

Wine, spirit, and ale, are all alike, as it regards the fact of their being stimulants; they only differ somewhat in kind and degree.

I shall speak, for the present, only of wine, for the sake of convenience. But whatever I shall say of wine, is to be considered as equally true of the others: and if what I have taught you, in my preceding Letters, be true, what I shall say now of stimulants must be true also.

If wine be productive of good, what is the nature and kind of good which it produces? Does it nourish the body? We know that it does not; for the life of any animal cannot be supported by it. Besides, if you have understood what I have said as to the nature, manner, and mechanism of nutrition, you will see at once, from the very mode in which the body is nourished, that whatever is capable of nourishing, must be susceptible of conversion into the solid matter of the body itself. But fluids taken into the stomach are not capable of being transmuted into solids, but pass off by the kidneys, as everybody knows.

If, indeed, the fluid drink contains solid matters suspended in it, then these solid matters can be assimilated to the solid body, and so are capable of nourishing it; as in the instance of broths, barley-water, &c., &c.; but the fluid in which these solid particles were suspended, must pass out of the body by the kidneys. Nothing, therefore, can contribute to the nourishment of the body which is not itself solid, or in which solids are not contained. Milk, the moment it reaches the stomach, is converted into curds and whey. The whey passes off by the kidneys—the solid curd nourishes the body.

If, then, it be said, that although wine is incapable of nourishing the body wholly and by itself alone, it may yet contain some nourishment, it is clear that this nourishment must depend upon whatever solid particles are suspended in it.

Now, if you evaporate a glass of wine on a shallow plate, whatever solid matter it contains will be left dry upon the plate; and this will be found to amount to about as much as may be laid on the extreme point of a penknife blade; and a portion—by no means all, but a portion of this solid matter, I will readily concede, is capable of nourishing the body—a portion which is about equal to the flour contained in a single grain of wheat.

But still I am entitled to ask, what good you propose to yourself by drinking wine? Because, if you really drink it for the sake of the nutriment it affords you, then, I say, why not eat a grain of wheat, instead of drinking a glass of wine; from which grain of wheat you would derive just as much nourishment as you would from the glass of wine? And by eating which, instead of drinking the wine, you would avoid the mischief which must necessarily be inflicted on your stomach by the deleterious spirit contained in the wine. Why go this expensive, and, as it were, roundabout way, in order to obtain so minute a portion of nutritious matter, which you might so much more readily obtain by other means—and without the necessarily accompanying evil inflicted by the spirit contained in the wine?

Wine, therefore, possesses no power to nourish the body; or, at least, in so minute a degree as to make it, as an article of nourishment, wholly unworthy notice.

Does it strengthen the body?—Let us see.

I have proved to you, in my former Letters, that health and strength depend upon a high degree of contractility: and I have proved, also, that a high degree of contractility can only exist when the body is rapidly and well nourished. Whatever, therefore, is capable of strengthening the body, must do so by increasing the contractility of its fibre: and whatever is capable of heightening contractility, must do so by a vigorous and rapid nutrition of the body. But we have just seen that wine possesses scarcely any nutritious virtues at all. How then can it strengthen the body? It cannot:—it is manifestly, demonstratively, glaringly impossible. But to nourish and strengthen it, are the only two good things which any kind of nutriment is capable of contributing to the body. I have just proved that wine possesses no power to effect either of them: it follows, therefore, as a direct necessity, that it is productive of no good at all.

Is wine certainly pernicious?

I have already proved that it is unnecessary: and it has ever been universally held, by medical philosophers, that whatever is unnecessary is detrimental. The simple fact then that wine is unnecessary, is a sufficient proof that it is injurious. Nor is the truth of this medical maxim at all wonderful. The finest hair introduced amongst the machinery of a watch is sufficient to derange its movements. And when one considers the exquisite delicacy of those properties on which life and health so mainly depend—I mean, contractility and sensibility, as well as that of the whole nervous system—one cannot certainly feel surprised that anything brought into contact with them, which is not strictly proper to them, should disorder the nicety of their delicate functions. But I have other proofs.

You will admit, at once, that the practice of drinking is followed by a high degree of morbid sensibility:—witness the nervous and tremulous anxiety of the *débauché* on the morning following a debauch. But I have long since shown you, that increased sensibility and vigorous contractility are incompatible; and that whatever augments sensibility, must have the effect of lowering contractility. But health and strength depend on vigorous contractility. If wine, therefore, heightens sensibility, it must diminish contractility; and thus, by impairing that property, impair the health and strength which depend on that property.

Again, if you will allow it to be true, that it is the sensibility of our organs which establishes the due relation between ourselves and external objects—teaching us what is good for us, and what injurious, by the pleasure or pain which the several external objects confer or inflict—then it again follows, *par nécessite*, that wine is hurtful; because wine, when tasted for the first time by unsophisticated palates, always impresses them disagreeably. To him who swallows a glass of raw spirit for the first time, the effects are painful to a high degree—almost suffocating. And no child would like wine or beer, unless taught to do so by precept, example, or habit.

Again: What is poison? Is it not any substance which, when taken into the system, has the effect of disordering some one or more of the actions which make up the sum of life; and which, if taken in sufficient quantity, will destroy life itself? This is the true definition of poison. Is it not, also, the strictly true definition of ardent spirit? Spirit has the effect of disordering the nervous system to so great a degree, as to produce intoxication; exciting the brain, sometimes to madness, always to folly, in an extraordinary manner. Is not this to disorder the functions of life? And if it be taken in sufficient quantity—if a man swallow a pint of overproof rum at a draught—will it not kill him? It will. Wherein, therefore, does spirit differ from poison? Only in the dose.

Aye, but (you may say) it is only poisonous when taken in sufficient quantity! True:—neither is prussic acid, neither is arsenic, neither is mercury, neither is opium. All these poisons are daily given by medical men, without destroying life. Why? Because they are not given in sufficient quantity. But will you, therefore, contend that they are not poisonous?

It is the effect of prussic acid to lower the nervous system below the natural standard. It is the effect of ardent spirit, first to excite the nervous system above, and then to depress it below, the natural standard also. Both these effects are poisonous—both will destroy life, if carried far enough: neither will destroy life, if not carried far enough. Prussic acid, therefore, and ardent spirit, are equal poisons: though neither will destroy life, unless taken in sufficient quantity. But would you willingly continue to swallow prussic acid daily, merely because you admire its delicious flavor, comforting yourself the while, by saying, that it could do you no harm, because you did not take it in sufficient quantity to destroy life? And, above all, would you do so, knowing it to be unnecessary?—Yet have I not proved that wine, spirit, and ale are unnecessary?

But if you be impenetrable to rational argument, you dare not deny the result of direct experiment. Observe:—"Sir Benjamin Brodie found, that by the administration of a large dose of alcohol (ardent spirit) to a rabbit, the pupils of the eyes became dilated, its extremities convulsed, and the respiration laborious; and that this latter function was gradually performed at longer and longer intervals: and that it, at length, entirely ceased. Two minutes after the apparent death of the animal, he opened the thorax (chest), and found the heart acting with moderate force and frequency," (now mark what follows!) "circulating dark-colored blood. The same phenomena resulted from the injection of two drops of the essential oil of bitter almonds" (the active principle of which is prussic acid), "diffused in half an ounce of water, into the bowels of a cat."*

Here, then, we have direct and irrefragable proof that ardent spirit is not only a poison, but a poison of the very same nature as prussic acid—producing the same effects—killing by the same means, viz., by paralyzing the muscles of respiration, and so preventing the necessary change of the black blood into vermilion blood;—about which black and vermilion blood I have said so much, in my Letters.

The strength (that is, the intoxicating power) of wine and ale depends upon the ardent spirit which they contain.

A great deal of mischief has arisen from the misapplication of the term "strength" to the intoxicating power of "strong drinks," as they are called. Potions are said to be "strong;" and thence, I have no doubt, first arose the silly notion that they possess the power of strengthening the body—of communicating some portion of their own strength, I suppose, to the body of the potator. People seem to suppose that by swallowing strong drinks they actually swallow strength as though strength were some tangible substance, which can be chewed, swallowed, and assimilated, like a potato. We say that onions have a "strong smell;" and we might as well expect to derive strength from smelling onions, as to do so by drinking fluids which have a strong flavor. We call them strong because they affect us strongly. And this, of itself, is another proof of their mischievous tendency;—for whatever affects us strongly, cannot be "chip-in-porridge;" and if it be not good and necessary, it must, of necessity, be not only simply injurious, but very highly so.

But, after all, my dear John, mankind in general know how to live, as well as I can tell them. They do not err from ignorance. They are spell-bound by passion—seduced by pleasure, and hood-winked;—but they are hood-winked willingly. They know that spirit, wine, ale, &c., are unnecessary, and even hurtful. All writers, in every age, have written in favor and praise of temperance, both in eating and drinking. Universal experience proves its necessity, if we would possess the "*mens sana in corpore sano.*" Individual experience proves it equally:—the horrible sensations felt in the morning after a debauch—the frequent necessity which most men have been under of desisting wholly from intoxicating drinks in order to recover their lost health—the utter loathing

* *Paris's Pharmacologia*, vol. 1, p. 244. Sixth Edition.

with which he who is not habitually a toper regards, next day, the beverage which intoxicated him—and fifty thousand other tokens, too clearly evident to be mistaken. The very word "intoxicate" signifies "to shoot with poisoned arrows." If men really thought that daily doses of wine, and spirit, and ale, were necessary to improve their health and strength, those who could afford it would give them to their favorite hunting horses and their pet dogs. Yet they do not this. The training jockey does not mix wine, or brandy, with the daily allowance of water to the horses he has under training for the course. All men know that luxurious feeding is injurious to health, and rigid temperance beneficial. All teachers have taught it, and all experience proves it. *Επει τι*, said Euripides, hundreds of years ago,

> *Επει τι δει βροτοισι πλην δυοι ν μοιοιν,*
> *Δημητρος ακτης, πωματος θ' ὑδρηχοου.*

That is: "What need has man of more than two things only—bread and water?" But the fact is, my dear John, the rogues like it, and will have it, right or wrong;—or, if they be blind, they are, at all events, determined not to be cured of their blindness. They had rather not see the evils they incur, than sacrifice the pleasure of incurring them. What they really want, is some rule which shall enable them to continue to enjoy the table and the bottle, and yet escape the consequent evils. They want a sort of impenetrable armor—a kind of philosopher's stone—some magic elixir, which shall confer on them a talismanic immunity from the evils of intemperance. They would fain discover some Styx, wherein to baptize themselves, and become invulnerable to disease. If a thousand men were to read this Letter, there probably would not be one but would see, and feel, and acknowledge, that its doctrines are true; but it is no less probable that every man of them would close it when he had done, and call for his brandy-and-water with as much composure as though he was doing the most sensible thing in the world. Or perhaps they would each remark: "Well! I have drunk brandy-and-water for these twenty years, and I do not see that it has done me any harm; so I shall e'en go on as heretofore." Yet, if an impertinent countryman insult him in the street, he must pocket the affront, and slink off, or suffer all the trouble and inconvenience consequent on sending him to the station-house, instead of quietly knocking him down where he stands. Why is this? Why! because his brandy, and wine, and luxurious habits, and full feeding, have rendered him no match for the hardy countryman. Yet he presumes to say, that his brandy-and-water has done him no harm, forsooth!—"I have drunk a gallon of beer every day," once boasted a certain hostler, "for the last thirty years, and I never was in better health than I am at this moment." The next day a fit of apoplexy laid him dead in the ditch.

But does there really exist any such philosopher's stone as I have mentioned above? Are there any means by which a man may enable himself to indulge freely in the pleasures of the table with impunity? I believe such means do exist—not of escaping with absolute impunity, but certainly with comparative impunity. And I believe, moreover, that I shall confer a more acceptable benefit by pointing out these means, than if I were to write a wagon-load of volumes, all crammed with dietetic rules from "title-page to colophon." But do not, my dear John, like the "profanum vulgus," despise the means which I shall point out to you, because of their simplicity. The world seldom attach much value to things which are plain and easily understood; only bestowing reverence on things which they can by no means understand—things complicate, mysterious, and incomprehensible. But be you wiser. The dervish, in the Eastern allegory, well aware of this weakness, knew that it would be in vain to recommend the sultan, for the cure of his disease, simply to take exercise. He knew that mankind in general require to be cheated, gulled, cajoled, even into doing that which is to benefit themselves. He did not, therefore, tell the sultan, who consulted him, to take exercise; but he said to him, "Here is a ball, which I have stuffed with certain rare, costly, and precious medicinal herbs." (If they had not been costly and precious, the sultan would have thought nothing of them.) "And here is a bat, the handle of which I have also stuffed with similar herbs. Your highness must take this bat, and with it beat about this ball, until you perspire very freely. You must do this every day." His highness did so; and, in a short time, the exercise of playing at bat and ball with the dervish cured his malady.

Now, my dear John, this same *exercise* which cured the sultan, is precisely the talisman which I am about to recommend to your adoption, as the chief means of remedying bad health, and of preserving that which is already good.

Ever yours,

E. Johnson.

LETTER XI.

Umberslade Hall, near Birmingham.

MY DEAR JOHN,

Before I enter into particulars, I beseech you to recollect what I have said to you in one of my former Letters; viz., that if you admitted the truth of what I then said, you would not afterwards be at liberty to dispute the truth of what I am about to say now; any more than he who admits the truth of the doctrines taught in the First Book of Euclid, can afterwards deny the truth of those taught in the Second;—because, if the one be true, the other, of necessity, must be true.

Before you proceed further, therefore, do me the favor to re-peruse carefully and attentively my Fourth and Fifth Letters. In doing this, pay particular attention to the definition of life—the manner in which it is supported; viz., by the perpetual wasting and regeneration of the body out of the blood—to the definition of health—to the description of the functions performed by the contractility and sensibility of our organs—to stimuli—to the uses of the circulation of the blood—to the different characters of the two sorts of blood contained in the body, &c., &c. By the way, all the subjects are not embraced in those three Letters; but, as it is absolutely necessary that all these should be well understood before you can clearly comprehend the full force of what I am now going to say, you had better carefully re-peruse the whole, before you proceed further.

Supposing, then, that you have done this, and done it understandingly; and supposing that you assent to the several definitions which I have given of life, health, nutrition, contractility, sensibility, stimuli, &c.; and supposing that you see no reason to doubt the accuracy of my statements relative to the offices performed by the absorbents, secreting glands, circulation of the blood, &c.; I now proceed to point out to you my reasons for recommending *exercise* as a talismanic agent in the prevention and cure of disease; entreating you always to remember, that by disease I here mean solely that depreciated state of the living actions —that sickly condition of the body—in which there is no structural lesion of the organs—in which no single organ is affected by any accidental disorganization, or defined and denominated disease —but in which all the nutritive actions are feebly performed, and in which the general tone of the health and strength is universally lowered: in a word, I mean that anomalous state of the health usually termed "*indigestion*," or "*dyspepsia*." And if I can teach you how to avoid this, I shall have taken a large stride towards teaching you how to escape almost all other disorders, especially chronic disorders: for, for the most part, it is general disorder which produces local disease, and not local disease which produces general disorder. But to proceed:—

Life, in the wide and physiological acceptation, consists of all the actions of which living beings are capable; not only the internal actions, as of the heart, vessels, &c., &c., but also of the external actions, as of the limbs in running, leaping, &c. But in a medical point of view, when speaking of life, the internal actions only are indicated—those invisible and inappreciable molecular motions which are constantly going on in the ultimate tissue of our organs, and by which nutrition is effected.

All physiologists agree that life consists in the constant wasting and reproduction of the body, particle by particle—by a perpetual analysis of the old particles composing our organs, and a perpetual synthesis of new particles derived from the blood—by a perpetual pulling down of the old materials, and a perpetual replacement of them by new—by perpetual disorganization, and perpetual reorganization.

The operation by which life is supported may be illustrated by the operation by which motion is supported and communicated by two cog-wheels acting on each other. Keep your eye steadily fixed upon the point at which the cogs of the two are interlocked. What do you observe? Why, that, at every instant, the empty space which is presented by one wheel, is instantly filled up by a tooth of the other wheel, to be almost instantly emptied again, and again re-filled. Thus it is, that at every point of the body, and at every instant, little empty spaces are made, which are immediately filled by the nutritive arteries, to be again emptied, and again filled.

Another postulate necessary to my forthcoming argument, and which is also indisputable, is this—that you cannot increase the size of your natural body, the substance of your natural fibre, by eating. This is certainly true. For if it were otherwise, the magnitude of the body would be equally enlarged at every point. If you increased its transverse diameter, you must also increase its longitudinal diameter. You could not make it broader without also making it longer. But this is contrary to known facts, for no man can make himself taller by eating; nor add, in the slightest degree, to the length of his fingers and toes. Yet the bones are nourished by the self-same food as the rest of the body, and by the same processes of assimilation.

By eating, therefore, you may superinduce fat over the body; but the magnitude of the solid body itself cannot be increased.

You will be pleased, too, to recollect, that I have already shown you (Third Letter), that energetic contractility can only reside in recently-organized matter; and that, therefore, rapid reorganization is absolutely essential to energetic contractility. And I have also, in one of the foregone Letters, proved to you, that all the living actions, external and internal, are performed by virtue of contractility; and that health and strength depend—absolutely depend—upon an energetic contractile power.

Now, then, observe the force and tendency of the following sorites.

Health and strength depend upon energetic contractility—

Energetic contractility depends upon rapid reorganization—

Rapid reorganization depends upon rapid disorganization—

Therefore, health and strength depend upon rapid disorganization.

The first process, therefore, in that chain of processes by which life is not only supported, but in which life really consists, is—what?—Eating?—No:—it is the wasting, the pulling down, the disorganization of the body. You must waste it, before you can nourish it. To the unreflective, this will seem paradoxical. Yet a moment's thought, without the parade of logic, should be sufficient to convince us of its truth. For does not appetite, in the natural order of things, precede the act of eating? And what is appetite, but a sensation warning us that the body has suffered waste, and calling upon us to repair it?

I say, that reorganization depends upon disorganization: because having shown that the body's fibre cannot be enlarged, it is clear that no new materials can be added until a corresponding portion of the old materials has been removed. It must, therefore, be pulled down, before it can be built up—impaired, before it can be repaired—disorganized, before it can be reorganized.

Now, the natural means by which the body is disorganized are, the exhalation from the lungs—perspiration from the skin—the several other excrements—and the formation of the several secretions required for the assimilation of our food—as, the gastric juice, bile, &c., &c.

You know how greatly bodily exertion augments perspiration, and increases the rapidity of breathing—and therefore, necessarily, the quantity of pulmonary halitus, or "breath," as it is called. Very well;—in like manner, also, it increases all the other secretions—those several fluids on which the assimilation of our food wholly, and solely, and absolutely depends.

Bodily exertion, therefore, promotes, and that most rapidly and powerfully, the disorganization of the body; and is, in fact, as far as I know, the only means of promoting it: as idleness is the infallible means of retarding it—that is, of retarding those processes, the activity of which are an absolute *sine-qua-non* to health and strength.

By a former syllogism, it has been proved that health and strength depend upon the rapid disorganization of the body: and I have just shown that rapid disorganization can only be effected by rapid exertion, or bodily labor. Hence legitimately arises another important syllogistic truth; thus:—

Health and strength depend on rapid disorganization—

Rapid disorganization depends on rapid exertion—

Therefore, health and strength depend, primarily, on rapid exertion.

From the whole, then, there results this general conclusion:—that there can be no such thing as perfect health and strength, without bodily exertion—that it is contrary to the very scheme of man's existence—that it is not in the nature of things—and that the philosophy of life and health, the light of science, the testimony of all ages, and the irresistible force of irrefutable argument, prove it to be impossible.

But there is another powerful argument, proving the necessity of bodily exertion. You must have observed, in reading my former Letters, that everything—no matter what—that everything which is done in the body, is done by virtue of the circulation of the blood. You must have remarked, that all the phenomena constituting life and health are effected, directly or indirectly, by the circulation;—that almost thought itself is the result of it:—most certainly the power of thinking is greatly modified by it.

Seeing, then, that the blood's fluxion is the all-efficient agent by which all the living phenomena are effected, it surely can require no great stretch of faith to feel convinced at once, that if this agent be allowed to doze at its post, infinite mischief must ensue; and that whatever is capable of keeping its energies in constant activity, is of the very highest value to the welfare of the system. And further, that whatever circumstances—such as sloth, and the other habits which I have enumerated as conducive to a languid circulation—whatever circumstances are calculated to lull its energies to repose, are in the highest degree detrimental. And the influence which bodily labor exercises over the circulation, everybody knows;—it is felt in every pulse of the body. Besides, the heart is a muscle, similar in its nature to the muscles of the arm or leg. Exercise, therefore, has the same influence in strengthening the heart (and, of course, through it, the circulation) as it has in strengthening the muscles of the leg or arm; and most men are well acquainted with its influence in this respect.

But there is yet another view of the subject, which I shall now open to you.

You know that our relation to external things is established by virtue of the sensibility of our organs, and that the degree of sensibility depends upon the degree of mobility of the nerves. I have also shown you how this mobility, and consequently our sensibility, is increased by a languid circulation, and how it is diminished by a vigorous one—by which the blood is driven energetically into all the capillary vessels, causing their coats to be well distended, so as to exert a constant and steady lateral pressure upon the nerves which run between them.

The circulation, therefore, is a resisting power—a power directly opposed to sensibility. And this resisting power may be carried so far as to produce absolute insensibility; as in those cases of apoplexy, in young, athletic, healthy men, which is the result of plethora—that is, of having too much blood, and too powerful a circulation.

The circulating and sensitive, therefore, are two antagonizing powers. And, as sensibility is the power by which we receive the consciousness of impressions, so the sanguineous circulation is the power by which we resist the consciousness of impressions.

Now, the same impressions will produce both pleasure and pain:

the difference being only in the force of the impressing cause. The same impressing cause which, slightly exerted, would yield pleasure, will, if its force be sufficiently increased, be productive of pain. This requires no illustration. But, to increase our impression-receiving power is the same thing as to increase the force of the impressing cause. By increasing, therefore, our sensibility, we virtually increase the force wherewith external objects impress us. And thus it is, that persons whose sensibility is morbidly acute, derive only pain from those causes from which the robust and healthy receive only pleasure. The natural relation between themselves and the external world is destroyed ; and they are living in a sphere for which they have become no longer fitted—with which they no longer possess the necessary and natural affinity. They are now "three-cornered men, thrust into round holes"—they do not fit their position. This morbid sensibility is a source of immeasurable calamity. To all it is the cause of continual irritation and painful sensation; to some it is an exhaustless fountain of misery: witness the lives of J. J. Rousseau, Gilbert, Zimmerman, Cowper, and numerous others; amongst whom, I think, I might mention the late Lord Byron.

But sensibility being the impression-receiving power, and the sanguineous circulation being the impression-resisting power, we have only to increase the latter, in order to diminish the former; and so restore the necessary and natural balance. And this brings me to the point at which I wished to arrive ; viz., bodily exertion; this being the only means I know of invigorating the power of the circulation.

This manner of considering the circulating and sensitive powers furnishes a ready answer to that hackneyed and silly question: "How is it that we see men arrive at a good old age, who have, all their lives, been drunkards ?"

Let me observe by the way, that these instances are extremely rare: and that they only appear to be frequent, because they are obtruded on our notice as remarkable occurrences. An instance of this kind is never allowed to escape our observation; because man is ever eager to catch at anything which may offer itself as an excuse for indulging in those habits to which his inclination urges him. Every such instance is, therefore, carefully registered; while the thousands who drop daily, like rotten sheep, into premature graves, the victims of intemperance, are neither minded nor marked. "He died," say they, "of this, or that, or the other disease;"—never stopping to inquire how that disease was incurred.

But the true reason why a few can commit habitual intemperance with comparative impunity is, because, in these persons, the impression-receiving power (sensibility) is naturally exceedingly dull, while the impression-resisting power (viz. the circulation) is naturally extremely vigorous. Their blood, propelled by a large and powerful heart, and rapidly and thoroughly oxidized by capacious lungs, is driven, with energetic force, to every point of the body; thus not only enabling it to resist the impressions of the deleterious matters introduced into the stomach, but also rapidly to repair whatever slight injuries are really inflicted.

There is yet another reason why bodily exercise is indispensable to health. The blood is wholly incapable of fulfilling any of its multifarious and all-important offices (except the secretion of bile) until it has been oxidized in the lungs. The more rapidly, then, that it is driven through the lungs, the larger will be the proportion of it which is oxidized, and so rendered fit to fulfil its function of nourishing the body—the greater will be the proportion contained in the arteries, where alone it is of use, and the smaller will be the quantity of black blood left in the veins, where it is of no use, except as before excepted.

I believe it is possible, by very rapid exertion, to fill almost every vein in the body with arterial blood. I have not room here, to detail the observations which have led me to this conclusion; but I do not speak unadvisedly. Nor would the secretion of bile be stopped by this state of things; for it has been proved, that although, under ordinary circumstances, it is secreted from venous blood, yet it can be secreted from arterial.

Now, I do not mean to say that it would be advisable for you to arterialize the whole of your blood. And there is no fear of it; for it would require greater exertion than any man would, or even could, voluntarily, undergo. But be assured of this, that the greater the quantity of arterial blood, and the less the quantity of black venous blood, contained in your body, the stronger, aye, and the happier and more light-hearted you will be—and the only means of arterializing the venous blood, is bodily exertion. The elasticity of mind and joyousness of heart, which *exercise* proverbially affords, are the direct results of an increase in the quantity of arterial, and a decrease in the quantity of venous, blood. The increase of animal spirits, as well as the increase of animal strength, must always correspond with the increase of arterial blood.

As we breathe for the purpose of oxidizing the black blood, then the oftener we are compelled to breathe, the better; because every time we breathe, a portion of black blood becomes oxidized, and fit for use. The increased rapidity of breathing consequent upon exertion is an increased rapidity in the function of oxidizing the blood—one of the most important of all the living actions. During exertion we drink, as it were, oxygen from the air. And this oxygen is the only stimulating drink which we can take, with advantage to ourselves, for the purpose of invigorating our strength, and elevating our animal spirits. It is the wine and spirit of life —the true *eau de vie*—with an abundance of which Nature has supplied us ready made—and it is the only one proper to man. If you be thirsty, drink water—if low spirited, drink oxygen. That is to say, take active exercise, during which you will inhale it.

Besides all this, every time the blood has completed its circle of circulation, a part of the great office of nutrition has been accomplished. The more rapidly the blood, therefore, is, by natural means, circulated through the body, the more rapidly does the process of nutrition go on.

You may compare the living actions to the actions of a hand-cornmill, the heart representing the first wheel, which puts into motion all the other wheels: and bodily exertion may represent the man who turns the crank, attached to the first wheel. Now, the more rapidly the man turns the crank, the more rapid will be the motion of the first, second, third, and all the other wheels; and the more rapidly will the corn be ground. At the same time, if the crank be turned with inconsiderate fury, the machinery may be deranged, and the mill broken.—So, bodily exertion is not to be furious. A horse may be ridden to death;—and, therefore, bodily exertion may be carried too far.

I have said, that persons of sedentary habits become frequently sensible of a feeling of want—a sinking at the stomach, as they express it; which they seek to relieve by eating or drinking. I have said, too, that although these persons require the excitement of a stimulus, yet food or wine does not furnish the stimulus required, but, on the contrary, only adds to the evil. What they want is oxygen.

You know I have all along mentioned four things as necessary to life; one of which, you cannot have forgotten, is *stimuli*. But

I shall disuse the word "stimuli;" because being used in the plural, it is awkward to introduce it correctly without periphrasis; and I will use the word "excitement" instead.

The exciting properties of arterial, that is, oxidized blood, I have just been describing to you, while showing you how rapid exercise produces its exhilarating effects; viz., by increasing the quantity of arterial blood, and driving it, in rapid currents, through all the countless avenues of the brain and body. It is to the lively leaping of the living current, pregnant as it is with the exhilarating wine of life—oxygen—that we owe all the bounding buoyancy, the elastic light-heartedness, which rapid motion and rapid exercise impart. In one of the volumes of Byron's works is the following note:—"A young French renegado confessed to Châteaubriand, that he never found himself alone, galloping in the desert, without a sensation almost approaching to rapture which was indescribable."—The circumstance of this man being alone in a desert had little to do with his rapturous sensations: he owed them to the rapid circulation and oxidation of his blood, produced as the joint effects of rapid exercise and rapid motion. The fox-hunter owes his pleasure to the same causes; and also the impunity with which he breakfasts on ale and brandy, and sleeps on half-a-dozen bottles of wine, and rises without a headache.

I cannot help inserting here a short extract from a very grateful letter I lately received from Stourbridge: "It is with gratitude," says the writer, "that I send these few lines to show that you have not written in vain. Before I read your work I had generally bad health—ever since, my health has been excellent, and I feel a happiness within me which I cannot describe. My employment is sedentary, 'chaffering behind a counter,' from seven in the morning till eight at night, yet I manage to get a run or a walk of seven or eight miles before business, and can walk fourteen miles before breakfast with ease, which gives me good appetite, good-humor, and good health."

Excitement, therefore, my dear John, is necessary; we cannot be healthy without it: and you and I only quarrel about the kind of excitement. This natural necessity for, and craving after, excitement is evinced in the numberless habits to which we addict ourselves, in order to obtain it. The habits of drinking and smoking owe their favor to the temporary excitement they afford. The reason why we crave after these unnatural kinds of excitement is, because we have lost a part of the excitement which is natural and necessary to us. It results from a languid and lazy circulation—a deficiency of oxygen—a gorged state of the venous system with black, devitalizing blood; and a deficiency of that stimulating and vivifying blood, whose color is vermilion, and which is proper to the arteries. Those distressing sensations of sinking, and want, and languor, and low-spiritedness, of which dyspeptics complain, accrue to them from the same causes. They are deficient in excitement—they want excitement; they want to have their brains, and heart, and whole system, stimulated, spurred, by the exciting properties of vermilion oxidized blood, driven merrily and forcefully to every point of the universal tissue.

We require a stimulant, then, certainly; but the only stimulant which will serve our purpose is arterial blood in energetic circulation: and the only means to procure this is bodily exertion. "*Exercitium naturæ dormientis stimulatio, membrorum solatium, morborum medela, fuga vitiorum, medicina languorum, destructio omnium malorum.*"—"Exercise is the awakener of dozing nature, the solace of the limbs, the healer of diseases, the chaser of vices, the medicine of listlessness, the destruction of all evils."

One word more for bodily exertion, as the means of increasing bodily strength;—and without health there cannot be strength.

Observe the manner in which horses are trained for the course. They are made to undergo more and more exertion, day by day, until the requisite degree of strength has been achieved. Reflect on this:—they strengthen these horses by making them daily undergo severe labor. They do not rest them, in order to strengthen them; they work them, in order to strengthen them. "Aye, but," says some wiseacre, "a horse is a horse, and a man is a man." Blockhead!—What then? We have but to exchange the race-course for the prize-ring, and the argument still remains in full force. Besides, have I not already shown that the actions which constitute the life, health, and strength of a horse, are precisely the same as those which constitute the life, health, and strength of man?

The prize-fighters will also furnish us with an example of the fact before stated; viz. that the high degree of contractility consequent upon an energetic circulation is hostile to, and incompatible with, much sensibility; these fellows becoming almost insensible to blows, unless dealt with an energy capable of felling an ox. They furnish an example, too, of another fact, which I have already stated (p. 85, in a note); viz., that well-filled arteries and a vigorous circulation are highly conducive—I believe absolutely necessary—to equable and amiable temper; for these men are generally remarkably easy and well-tempered fellows. On the contrary, if you seek a perfect example of pettish, irritable, quarrelsome, unforgiving, querulous, snappish, cat-like, unsoothable, spiteful, and sulky temper, you will find it in the Spitalfields' weaver—the poor, dyspeptic weaver, "cabin'd, cribbed, confined," and cramped in his loom for sixteen hours a-day, in a room ten feet square; whose utmost exertion is, to throw a shuttle, four ounces in weight, backwards and forwards, about the length of his arm; and whose longest peregrination is from his own cabin to the counter of the gin-shop, and from the counter of the gin-shop to the door of his own cabin.

The fortitude of the Indian at the stake arises from the same circumstances of a highly energetic circulation. From his habits of life, his circulation is always vigorous, and his sensibility obtuse; but at the moment of torture, its energy is still further augmented, and his sensibility still further blunted, by the enthusiasm and exultation which he feels in maintaining the honor of his tribe, and in disappointing his enemies, who, he knows, are eagerly watching for any symptom of wincing. His circulation in impetuosity resembles a spring-tide; and his body becomes almost insensible to pain.

Again, when the circulation through the brain is highly excited by intense thought, the nerves arising from the brain become almost insensible to the impressions natural to them;—the ear hears not; the eye sees not; the olfactory nerves take no cognizance of odors.

SLEEP.

Solidification—that is, the conversion of blood into the solid parts of the body—goes on only during sleep. The chief end, indeed, and object and intention of sleep, would seem to be this final assimilation of our food—this solidification of the blood into the several solid parts of the body.

The accomplishment of this miraculous change seems to have required the perfect concentration of all the energies of the system upon itself. It seems to have required, that the attention (if I may so speak) of the brain and nervous system should not be distracted by any other object. It seems to have required that everything, both within and without the body, should be hushed into profound repose during the accomplishment of this nightly wonder, in order that nothing might disturb or interfere with the exquisite and miraculous processes employed to effect it. To this end, the portals of sensation are closed—the eye sees not—the ear hears not—the skin feels not—the very breathing is scarcely audible—the pulsations of the heart are scarcely perceptible: all the living energies are now concentrated into the greatest possible intensity, like rays of light into a focus; and directed, with almost complete exclusiveness, towards this simple object.

In the day, therefore, we make blood:—in the night, that blood is converted into solid matter. In the day, we garner up the building materials;—in the night, we repair the building. The hour of rising, therefore, ought to be the time at which our physical strength is at the greatest;—and with perfectly healthy persons this is the case. The languor which sickly persons feel in the morning arises from the process of repair not having been fully accomplished: the building has not been repaired, and therefore its strength has not been restored. The apparent additional strength which is felt, during the day, after eating, is only apparent;—it is merely excitement derived from the stimulus of food; in the first instance in the stomach: and after that food has been assimilated, of new blood in the system.

From all this, we learn two important truths: first, that we should take our severest exercise in the early part of the day: secondly, we learn how and why it is that late suppers are improper.—When you retire to bed with a full stomach, before the process of solidification can commence, the food which the stomach contains must be assimilated. The two operations of solidification and assimilation of food into blood cannot go on together; because, as I have just shown you, the process of solidification requires the concentration of *all* the living energies for its accomplishment. The commencement of this process, therefore, must be postponed until the assimilation of the supper to blood has been completed. But all the living energies, except that of solidification, are diminished in intensity during sleep. The secretion of the gastric and other juices, therefore, necessary for the assimilation of the supper to blood, will go on but slowly, and the completion of the process will be exceedingly protracted; and thus, so much of the season of sleep will be employed in the assimilation of food, that a sufficient portion of it will not be left for the solidification of blood. But this is not all the mischief; for the process of assimilation of the supper into blood has not only abstracted from the process of solidification a portion of that season (the season of sleep) which ought to have been exclusively devoted to its own accomplishment, but it has also robbed it of a portion of those living energies, the whole of which were due to itself; viz., that portion which has been consumed in the secretion of those juices necessary for the conversion of the supper into blood. When, therefore, the hour of rising arrives, it finds the body still unrepaired and unrefreshed; and the individual still overpowered with sleep, and disinclined to rise.

To conclude:—If you would preserve your health, therefore, exercise, severe exercise—proportioned, however, to your strength—is the only means which can avail you. Recollect, the body must be disorganized, wasted, sweated, before it can be nourished;—recollect the tale of the Dervish and the Sultan;—recollect the mode of training horses for the course, and men for the prize-ring. With plentiful bodily exertion, you can scarcely be ill;—without bodily exertion, you cannot possibly be well. By "well," I mean the enjoying as much strength as your system is capable of; and if you are in search of some charm, some talisman, which will enable you to indulge considerably in the pleasures of the table with comparative impunity, you will find it in bodily exertion, and in bodily exertion only.

But by exertion or exercise, I do not mean the petty affair of a three-miles' walk: I mean what I say—bodily exertion, to the extent of quickened breathing and sensible perspiration, kept up for three or four hours out of the twenty-four;—say, four or five miles before breakfast, four or five before dinner, four or five early in the evening; or to save the evening for other purposes, a healthy man may walk ten or a dozen miles before breakfast, with an advantage to himself which will, in a week or two, perfectly astonish him. Most men, even the operative manufacturers and shopkeepers, may do this, if they will take the trouble to rise early enough;—and, fortunately, the exercise taken before breakfast is worth all that can be taken afterwards.

It would be easy to show, that the health and strength of the mind is as much under the control of the circulation as is the health and strength of the body.—But I have already exceeded my limits.

Rules of diet, therefore, are of little use; and that little, only to those who cannot take the necessary degree of bodily exercise. The stomach of a healthy man will dissolve polished steel of the finest temper. What difference can it make to such an organ, whether it receives roast or boiled meat, eggs, oysters, cheese, butter, bread, or potato? and whether these articles have been thoroughly and minutely broken down by the teeth, or only imperfectly so? Sir Richard Jebb, when his patients asked him what diet they should use, was in the habit of replying, "Why, my dear madam, do n't eat the fender and fire-irons, because they are decidedly unwholesome; but of any other dish you may freely partake."

But, to those who, from any cause, cannot take bodily exertion, some attention to diet may be necessary. Even here simplicity and quantity, rather than quality, form the grand consideration. They cannot well take too little food; and wine and other strong drinks are wholly inadmissible. And let them only reflect on the mechanism of nutrition; on the manner in which our food nourishes us—what becomes of it after we have eaten it; and they cannot but clearly see that this advice is sound and wholesome doctrine.

Authorities without proof are of little value; otherwise, I could quote them in abundance, from all sorts of authors, in all ages of the world. But if you will not believe the evidence of such arguments as I have already adduced, neither would authorities convince you, though their name were Legion. I shall, however, conclude this Letter with four.

"The pith of nearly all that has been written on *hygiene*," says Dr. J. Johnson, "or the prevention of diseases—and of the Protean disorder among the rest—might be included under two heads, almost in two words—*temperance* and *exercise*." Again: "Disorders of the body, in these days, are engendered and propagated, to a frightful extent, by moral commotions and anxieties of the mind. And if I have proved that *corporeal exertion*, especially when aided by any intellectual excitement or pursuit, can obviate the evils that ensue to soul and body from these causes, I shall do some service to the community. The principle in question is neither Utopian, nor of difficult application: it is within the reach of high and low—rich and poor—the learned and unlettered. Let moral ills overtake any of these, and he is in the high way to physical illness. To prevent the corporeal malady, and to diminish, as much as possible, the mental affliction itself, the individual must tread in the steps—*haud passibus æquis*—of Xenophon and Byron. He must *keep the body active and the stomach empty.*"

"Exercise," says Hakesworth, "gives health, vigor, and cheerfulness, sound sleep, and a keen appetite. The effects of sedentary thoughtfulness are, diseases that embitter and shorten life—interrupted rest—tasteless meals—perpetual languor—and ceaseless anxiety."

"Temperance," says Burton, "is a bridle of gold; and he who can use it aright, is liker a god than a man."

But—I beg your pardon—I must make another short quotation, which has this moment occurred to me:—one which, though exceedingly short, embodies in itself the truth and wisdom of a hundred volumes: it is the following brief aphorism of the late Mr. Abernethy: with which I shall conclude:—"If you would be well, live upon sixpence a day, and earn it."

I am, my dear John,

Yours very truly,

E. JOHNSON.

APPENDIX.

Since the publication of the eighth edition of this work, a new mode of treating certain diseases has been introduced into this country under the name of Hydropathy.

Shortly after its introduction, I received several letters, wherein it was stated, that if the opinions expressed in my work on Life, Health, and Disease were sincere, they (the writers of these letters) thought I was bound to advocate the Hydropathic treatment, because, they said, the *principles* and doctrines involved in that treatment are identical with the *principles* and doctrines taught by me in that work. I also received, shortly after the receipt of these letters, a note from Captain Claridge, requesting leave to call on me. "For," said he, in his note to me, "I have just been reading your work on Life, Health, and Disease, from which I perceive that you are a *ready-made Hydropathist.*" After my conversation with Captain Claridge, I immediately started for Græfenberg, where I spent one entire winter. During my residence there, I had the most ample opportunities of observing the effects of the treatment as practised by Priessnitz. I soon became most thoroughly convinced that the Hydropathic method, though not capable of curing all diseases as some visionary enthusiasts believed, nor of entirely taking the place of medicine and the lancet, to the total exclusion of both; yet that, as a whole, it formed a system infinitely superior to all other systems of cure, and more generally applicable, safe, and successful. I found, too, that it was particularly useful in precisely those cases in which drugs were avowedly of no use at all—as chronic rheumatism, chronic debility, spinal weakness, indigestion, constipation, consumptive diathesis, and a whole host of nervous affections, head diseases, and irregular and undenominated ailments which defy classification. I returned home determined to advocate and adopt it—and I thank God that I have hitherto been enabled to do so with a success beyond my most sanguine expectations.

The chief object of this work has been to recommend exercise, and a more free exposure of our persons to the influences of the weather, as remedial and conservative measures of high value. And I have explained how these operate upon the body—viz. by accelerating the *disorganization* and *reorganization* of the body—by *more rapidly expelling the old and worn-out material, and supplying its place by new.* Now this is precisely the mode in which the Hydropathic treatment operates, as demonstrated by Liebig in his great work on Animal Chemistry,—and as also shown by me in my "Theory and Principles of Hydropathy." Frequent cold bathing and sweating have precisely the same effect on the body as exercise. They accelerate what Liebig calls the CHANGE OF MATTER—that is, the daily disorganization and reorganization of the elements of the blood and vital organs. But, in addition to frequent cold bathing and sweating, Exercise, and almost constant exposure to the weather, form most important elements in the Hydropathic treatment.

Not to have adopted, therefore, this treatment, would have been, in my case, to become a renegade from my own principles and doctrines.

As practised by Priessnitz, however, there is much in the treatment which is unscientific, and something which is both absurd and dangerous—at all events, to English constitutions—which differ widely from the German. I do not say this to detract aught from his merit, which is undoubtedly great. But I presume it will not be denied by any one that, had Priessnitz had the advantage of a medical education, or indeed any education at all, his treatment would have been much more perfect, and his success much more uniform and certain.

There is no part of that treatment which is wholly new. The wet sheet, wet body-bandages, sweating, cold bathing, exercise, and free exposure to the weather, have each, in its turn, been recommended by various writers and practitioners before. The great merit of Priessnitz consists in his having accumulated all these various remedial measures into one great system, bringing them all simultaneously to bear upon the patient as one treatment—which treatment must (in chronic cases), and *for the time being*, constitute the business of the patient's life.

There is another point on which the Hydropathic treatment has a strong analogy with the doctrines taught in this work—I mean its total rejection of all Alcoholic Stimulants as beverages.

As might have been expected, the Hydropathic treatment, on its first introduction, met with the most violent opposition on the part of medical men. One physician to one of our largest Metropolitan Hospitals gave a lecture to his pupils, which lecture was expressly devoted to instructing them in what they were to say, and how they were to act, should any of their patients, when they got into practice, consult them as to the propriety of trying this new treatment. They were not to run it down—not to ridicule it—because that, said he, would be the certain way to induce them to resort to it. On the contrary, they were to speak of it with respect—to acknowledge that it was useful in certain cases, and certain constitutions—but always to conclude by assuring the patient that in *his* or *her* particular case it would be highly improper and extremely dangerous. And this is the ruse—this the quirk—practised by the great bulk of medical men up to the present moment. They have ceased to rail at it—ceased to laugh at it. They admit that they sometimes recommend it.

But they always conclude by assuring the patient that *his* own case is one in which it is totally inadmissible.

I am rejoiced to perceive however that many among the more intelligent and higher standing in the profession are beginning not only to acknowledge its value, but even to introduce it into their practice. In the last number of the British and Foreign Medical Quarterly, Professor Forbes (its Editor) has a long article, written with the most manly frankness, by himself, on the subject of the Hydropathic treatment. In this article he has, in the most unequivocal language, boldly declared his conviction that it is a treatment of great power and value in a multitude of cases, both chronic and acute—and moreover that it is not a whit more dangerous than the ordinary drug practice, in the hands of properly educated medical men. "If it shall appear, however," says Professor Forbes, "*as we believe it will*, on further examination, that the external application of cold water is capable of being beneficially applied in the cure of diseases in *modes of greater efficacy*, and to a *much greater extent*, than has been hitherto practised by medical men, there remains only one course for the members of the profession to pursue, viz. to *adopt the improvements*."

Again: "Let us then consider a little further the consequence of repeated applications of cold, supposing, for the sake of argument, it is used with due reference to the constitutional powers, so as to create an increased activity of the vital functions. It appears to us that this *is exactly the thing needed* in the treatment of a *great many cases of chronic ailments*."

"In conclusion, we will venture to place on record the following, as *among* the more important impressions which have remained on our mind after a careful examination of the whole subject. In a large proportion of cases of gout and rheumatism, the water-cure seems to be extremely efficacious. After the evidence in its favor, accessible to everybody, we think medical men can hardly be justified in omitting, in a certain proportion of cases, at least a full trial of it. No evidence exists of any special risk from the water practice in such cases."

"In that very large class of complex disease, usually known under the name of chronic dyspepsia, in which other modes of treatment have failed, or been only partially successful, the practice of Priessnitz is well deserving of trial."

"In many chronic nervous affections and general debility, we should anticipate great benefit from this system."

"In chronic diarrhœa, dysentery, hæmorrhoids, the sitz-bath appears to be frequently an effectual remedy."

"We find nothing to forbid the cautious use of drugs in combination with Hydropathic measures. On the contrary, we are convinced that a judicious combination of the two is the best means of obtaining the full benefit of each. The water-cure contains no substitute for the lancet, active purging, and many other means necessary for the relief of sudden and dangerous local maladies. The banishment of drugs from his practice was necessary, and perhaps natural, on the part of Priessnitz: *the like proceeding on the part of qualified medical men, superintending water establishments in this country, evinces ignorance, or charlatancy, or both.*"

"With careful and discreet management, in the hands of a properly qualified medical practitioner, the water-cure is very rarely attended with danger."

"Notwithstanding the success of the founder of Hydropathy (Priessnitz), its practice by non-professional persons can neither be fully advantageous nor safe."

"Many advantages would result from the subject being taken up by the medical profession. The evils and dangers of quackery would at once be removed from it. Its real merits would soon be known. The tonic portion of its measures might then be employed in conjunction with special remedies of more activity, which, no doubt, would often prove exceedingly beneficial."

"Finally, it must always be remembered that the distinction between quacks and respectable practitioners is one, *not so much of remedies used*, as of skill and honesty in using them. Therefore, let our orthodox brethren be especially anxious to establish and to widen, as far as possible, this distinction between themselves and all spurious pretenders."

I need hardly add that I quite agree with these conclusions of Professor Forbes; especially that Hydropathy can never operate to the total exclusion of either drugs or the lancet—although undoubtedly it will diminish, to a very important extent, the *frequency* of the use of both—and that it cannot be either safely or advantageously practised by non-professional persons.

THE END.

JOHN WILEY'S LIST

OF

LATE PUBLICATIONS.

Alexander, Prof. J. A.

THE EARLIER AND LATER PROPHECIES OF ISAIAH. In 2 Vols. Roy. 8vo. cloth, $5 50.

"The sound, independent judgment which Prof. Alexander everywhere displays, combined with true candor, modesty, and a spirit of profound reverence for the inspired volume, distinguishes his work most advantageously from most of the critical productions of the age, and entitles it to be regarded as a model of Biblical investigation."—*London Patriot.*

"A rich contribution of philological exposition for the use of the clergy."—*Presbyterian.*

Alexander.—The Later Prophecies of Isaiah.

Royal 8vo. cloth, $2 50.

"A commentary of higher aim than the unfolding of a poem, and of profounder character than a mere repository of suggestive practical thoughts."—*N. Y. Recorder.*

Amber Witch, The; Or, Mary Schweidler.

The most interesting Trial for Witchcraft ever known. Edited by Dr. Meinhold. Translated from the German, by Lady Duff Gordon. 12mo. cloth, 50 cts.

"A beautiful fiction, worthy of De Foe."—*Quarterly Review.*

"The critics of Dr. Meinhold were fairly taken in, by his art stepping so exactly in the footprints of nature."—*Dem Review.*

Armstrong, Major-General.

NOTICES OF THE WAR OF 1812. In 2 vols. 12mo. cloth, $1 50.

Atheists—Voltaire and Rousseau against them;

Or, Essays and Detached Passages from those Writers, in Relation to the Being and Attributes of God. Selected and Translated by J. Akerly. 12mo. cloth, 50 cts.

Bartlett, Prof. W. H. C.—Optics.

AN ELEMENTARY TREATISE ON OPTICS, Designed for the Use of the Cadets of the U. S. Military Academy. Containing 8 folding plates. 8vo. cloth, $2.

Beckford, Wm., Author of "Vathek."

ITALY, SPAIN, AND PORTUGAL; With an Excursion to the Monasteries of Alcobaça and Batalha. 2 parts in 1 vol. 12mo. cloth, $1 25.

"Pleasing and picturesque as the clime and places visited, this is just the book for the indulgence of the '*dolce far niente.*'"—*London Lit. Gaz.*

"Rich in scenes of beauty and of life."—*Lond. Athenæum.*

Beecher, Revd. Edward, D.D. Baptism;

With reference to its Import and Modes. In 1 vol. 12mo. cloth, $1 25.

"The spirit of the scholar and Christian is conspicuous throughout; and in its candid, thorough manner, its excellent feeling, and its able argumentation, it strikes us as more nearly realizing the *beau ideal* of doctrinal controversy than is generally found."—*N. Y. Evangelist.*

Bible.

PATRICK, LOWTH, ARNALD, WHITBY, AND LOWMAN. A Critical Commentary and Paraphrase on the Old and New Testaments, and the Apocrypha. A new edition, with the text printed at large, in 4 thick vols. royal 8vo cloth, $15; sheep, $16 50.

⁎ Numerous testimonials have been obtained from the most learned divines of various denominations in this country, respecting the value of these expositions.

Bibliotheca Sacra and Theological Review.

Conducted by Profs. Edwards, Park, Robinson, and Stuart. Published quarterly, $1 each number, yearly subscription $4. Vols. 1, 2, 3, and 4, in cloth binding and back numbers, will be furnished at a discount of 25 per cent. from the retail price. Vol. 5, 8vo. cloth, $4.

"This Review continues to sustain its high character as the ablest periodical in the language, in theological and Biblical literature."

Blanchard, Laman.—Sketches from Life.

Edited, with a Memoir of the Author, by Bulwer. 12mo. cloth, $1.

"Amusement and instruction conveyed in a style, the beauty of which has not been surpassed by any of the writer's contemporaries."—*Dickens's News.*

"The memoir is one of the most affecting pieces of biography that we have ever read, and is alike honorable to the subject and the author."—*New Bedford Mercury.*

Bradford, A. W.—American Antiquities.

And Researches into the Origin and History of the Red Race. 8vo. cloth, $1 25.

⁎ A philosophical and elaborate investigation of a subject which has excited much attention. This able work is a very desirable companion to those of Stephens and others on the Ruins of Central America.

Bradford, Wm. J. A.—Notes on the North West;

Or, Valley of the Upper Mississippi: comprising the country between Lakes Superior and Michigan, East; the Illinois and Missouri Rivers, and the Northern Boundary of the United States; including Iowa, Wisconsin, &c. 12mo. cloth, $1.

"The work is divided into five parts. The first and second relate to the physical geography and history; the third, to the population, political system, civil divisions, municipalities, and topography; the fourth, to society, laws, pursuits, life, habits, and health of the Northwest, and to the public lands; and the fifth, and last, to the Aborigines and the monuments."—*Hunt.*

"A valuable addition to the library, and of great importance to the emigrants to the far West."—*Evening Mirror.*

Bull.—Hints to Mothers,

For the Management of Health during the Period of Pregnancy, and in the Lying-in Room; with an Exposure of Popular Errors in connexion with those subjects. 4th edition, 18mo. cloth, 50 cts.

Burke, Wm.—Mineral Springs of Virginia.

With Remarks on their Use, and the Diseases to which they are Applicable Second edition, enlarged, with a notice of Fauquier White Sulphur Springs, and a Map. 18mo. cloth, 75 cts.

Burton.—The Anatomy of Melancholy.

What It Is, with all the Kinds, Causes, Symptoms, Prognostics, and several Cures of It. In three Partitions, with several Sections, Numbers, and Sub-Sections—Philosophically, Medicinally, Historically Opened and Cut Up. By Democritus Junior. With Satyrical Preface. New edition, with Translations of Classical Extracts, and an Account of the Author, having a fac-simile title to the original edition. Royal 8vo. cloth, $2 50.

*** This is the best edition in one volume that is published.

"Contains more solid information than any twenty works ever compiled in the English Language."—*Byron.*

"A mine of forgotten scholarship."—*Dem. Review.*

Butler, Mrs. Fanny Kemble.

A YEAR OF CONSOLATION. 2 vols. in 1. 12mo. cloth, 75 cts.

"It contains the journal of a travel through Europe, and a residence in Italy; and is one of the pleasantest and most interesting books of the season."—*Courier.*

Campbell's Poems.

THE POETICAL WORKS OF THOMAS CAMPBELL. With a Steel Portrait. 32mo. cloth, 50 cts., or morocco extra, $1 25.

Cellini, Benvenuto, a Florentine Artist.

AUTOBIOGRAPHY OF. Containing a variety of information respecting the Arts and the History of the Sixteenth Century; with Notes and Observations of G. P. Carpani. Translated by Thos. Roscoe, Esq. 2 vols. in one, 12mo. cloth, $1 25.

"Rich in the most curious incidents, presenting a fine picture of Italian Life, Manners, and Art, and is invested with an interest truly romantic and dramatic."

"Cellini was one of the most extraordinary men in an extraordinary age; his life written by himself, is more amusing than any novel I know."—*H. Walpole.*

Chambers—Cyclopædia of English Literature.

A History, Critical and Biographical, of British Authors, from the earliest to the present Times. Edited by Robert Chambers. Complete in 2 vols. royal 8vo. $5, bound in full neat cloth.

—— Information for the People.

Conducted by William and Robert Chambers, Editors of Chambers's Edinburgh Journal, Educational Course, &c. New and improved Edition, in 2 vols. royal 8vo. $5, full cloth.

"A comprehensive poor man's Cyclopædia, and perhaps the most striking example yet given of the powers of the press in diffusing useful knowledge."

—— Miscellany of Useful and Entertaining Tracts.

20 vols. 12mo. bound in 10, $7 50, full neat cloth.

Cheever, Rev. George B.

WANDERINGS OF A PILGRIM IN THE ALPS. Part I.—In the Shadow of Mont Blanc. Part II.—In the Shadow of the Jungfrau Alps. By Geo. B. Cheever, D.D. In 1 vol. cloth, $1.

"Highly picturesque in its details, and written with a religious eloquence which becomes the profession of the writer."—*Simms' Mag.*

"There is an exuberance of fancy, a peculiar glow, in the language of Dr. Cheever, which gives a charm to his written productions possessed by few."—*Com. Adv.*

—— THE PILGRIM IN THE SHADOW OF MONT BLANC. In 1 vol. 12mo. cloth, 50 cts.

—— THE PILGRIM IN THE SHADOW OF THE JUNGFRAU. In 1 vol. 12mo. cloth, 50 cts.

—— LECTURES ON THE PILGRIM'S PROGRESS, and on the Life and Times of John Bunyan. A new edition in 12mo. size, containing over 500 pages. Uniform with the "Pilgrim in the Alps," cloth, $1

"We know of nothing in American literature more likely to be interesting and useful than these lectures. The beauty and force of their imagery, the poetic brilliancy of their descriptions, the correctness of their sentiments, and the excellent spirit which pervades them, must make their perusal a feast to all of the religious community."—*Tribune.*

—— THE JOURNAL OF THE PILGRIMS AT PLYMOUTH, in 1620. Reprinted from the original volume. With historical and local illustrations of Providences, Principles, and Persons. By Rev. Geo. B. Cheever, D.D. In 1 vol. 12mo. uniform with the above, cloth, $1; cloth gilt, $1 50.

"This reprint is really a treat to those who love the memory of their forefathers ..... The illustrations by Dr. Cheever, are exquisitely beautiful in conception and execution."

"The book is full of striking passages which we should love to copy."

Cheever, Rev. George B.—Continued.

——— MISCELLANIES: including "Deacon Giles's Distillery," &c. In 1 vol. 12mo. uniform with the above, cloth with Portrait, $1.

——— A DEFENCE OF CAPITAL PUNISHMENT. New edition, in 1 vol. 12mo. cloth, uniform with the above, 50 cts.

CHEEVER AND LEWIS ON CAPITAL PUNISHMENT. A defence of Capital Punishment, by Rev. G. B. Cheever; and An Essay on the Ground and Reason of Punishment, with special reference to the Penalty of Death. By Prof. Tayler Lewis. In 1 vol. 12mo. cloth, $1.

Church, Professor, in U. S. Military Academy.

CALCULUS. Elements of the Differential and Integral Calculus. 8vo. in sheep or cloth, $1 50.

Clarke, Mrs. Cowden.

CONCORDANCE TO SHAKSPEARE. A complete Verbal Index to all the Passages in the Dramatic Works of the Poet. One large vol. 8vo. cloth, $6.

*** The London price is $12, but having bought a large number of copies the American publisher can afford, at this very low rate, a work which cost the author twelve years' labor.

"A most surprising monument of the compiler's labor and enthusiasm . . . . . . incomparably the most valuable effort of the kind that has ever been given to the world."—*London Examiner.*

Clarke, Mrs.—Shakspeare Proverbs;

Or, the Wise Saws of our Wisest Poet, collected into a modern instance. Square 18mo. with Woodcuts, one neat vol. cloth, 75 cts.

"The present is the best collection of passages culled from Shakspeare which we know of, since it contains only those things which can be detached from their context without suffering any diminution of their effect."—*Evening Post.*

Coleridge and Southey.

COTTLE'S REMINISCENCES of S. T. Coleridge and Robert Southey. 12mo. cloth, $1.

"It affords the reader complete insight into the private history of two of the most remarkable men of the present era."—*Home Jour.*

"Mr. Cottle has unconsciously erected a monument to his own gentle goodness in these volumes, which well deserve a place in any library."—*Chtn. Enquirer.*

"The letters of Coleridge and Southey in this volume give us a very clear notion of the career of those authors."

"One does not know where to stop in the perusal of such a curious book as this."—*Evergreen.*

Cortez.—Despatches.

Addressed to the Emperor Charles V. written during the Conquest, and containing a narrative of its events translated with introduction and notes, by Hon. Geo. Folsom. 16mo. bds. $1 50. The same, larger edition, 8vo. bds. $2.

"This book has all the interest of a novel, and all the value of a history. What higher praise can we bestow upon such a work? This marvellous, this unparalleled story."—*Tablet.*

Cruikshank.—The Bottle.

In 8 plates. Price 25 cts.

"The immortal George is very great in this work. It is an old story, but told with such force and reality, as in the present day belong to Cruikshank alone. In a series of eight plates we have the career of the bottle-victim from the first drop to the maniac's cell, traced through every stage of misery, degradation, and crime."—*London Atlas.*

——— The Drunkard's Children.

In 8 Plates. Price 25 cts.

Dana, Prof. J. D.—A System of Mineralogy,

Comprising the most recent discoveries: with numerous woodcuts, and four copperplates. 2d edition, 8vo. cloth, $3 50.

"This work does great honor to America, and should make us blush for the neglect in England of an important and interesting science."—*Lond. Athenæum.*

"Bears marks on every page of great industry and determination in collecting the most recent facts. In completeness, systematic arrangement, and accuracy, it is believed to be exceeded by no other work extant."

De Hart, Capt. W. C.—Courts Martial.

Observations on Military Law, and the Constitution and Practice of Courts Martial; with a summary of the law of Evidence, as applicable to Military Trials; adapted to the Laws, Regulations, and Customs of the Army and Navy of the U. S. 8vo. law sheep, $3.

Diary of Lady Willoughby.

So much as relates to her Domestic History, to the eventful period of the reign of Charles the First, the stirring events of the Protectorate and the Restoration. 1st and 2d series in 1 vol. 12mo. cloth, 50 cents; or in extra cloth, gilt, $1.

"The simple, but antique style of language in which it is clothed, together with much that is beautiful in thought and expression, and an exquisitely well drawn picture of domestic life among those of rank and consequence in olden time, stamps the work with a novelty and interest which is quite rare."

"As a picture of the times, it is very true and graphic."—*Home Journal.*

Dickens—Carols, Chimes, and Cricket.

In 1 vol. 12mo. cloth, 75 cents, or fancy cloth, $1.

"On these three holiday tales, the reading world has already passed its verdict. Mr. Dickens, as a sketcher of every-day life, has probably no equal in the world. Perhaps no writer, not even Scott, has so many admirers of his works."—*Island City.*

Dickens—Christmas Stories.

THE CHIMES.—A goblin Story of some bells that rang an Old Year out and a New Year in.

THE CRICKET ON THE HEARTH.—A Fairy Tale of Home.

www.ingramcontent.com/pod-product-compliance
Lightning Source LLC
LaVergne TN
LVHW011231110826
845150LV00006B/1604

9781425514891